Nanotechnology and the Environment

Edited by Mousumi Sen

Published in London, United Kingdom

IntechOpen

Supporting open minds since 2005

Nanotechnology and the Environment
http://dx.doi.org/10.5772/intechopen.87903
Edited by Mousumi Sen

Contributors
Ali Salman, Mousumi Sen, Omar Messaoudi, Mourad Bendahou, Mostafa Esmaeili Shayan, Farzaneh Ghasemzadeh, Salah Abdelbary, Christos Argirusis, Pavlos Pandis, Georgia Sourkouni, Christos Vaitsis, Eirini Kanellou, Nikolaos Argirusis, Antonis Zorpas, Maria Mechili, Navin Kumar Mogha, Muthuraman Yuvaraj, Hadeer Abdelfattah, Kizhaeral Sevathapandian Subramanian

First published in London, United Kingdom, 2020 by IntechOpen
IntechOpen is the global imprint of INTECHOPEN LIMITED, registered in England and Wales, registration number: 11086078, 5 Princes Gate Court, London, SW7 2QJ, United Kingdom
Printed in Croatia

British Library Cataloguing-in-Publication Data
A catalogue record for this book is available from the British Library

Additional hard and PDF copies can be obtained from orders@intechopen.com

Nanotechnology and the Environment
Edited by Mousumi Sen
p. cm.
Print ISBN 978-1-78985-228-8
Online ISBN 978-1-78985-671-2
eBook (PDF) ISBN 978-1-78985-672-9

For EU product safety concerns:
IN TECH d.o.o., Prolaz Marije Krucifikse Kozulić 3, 51000 Rijeka, Croatia, info@intechopen.com or visit our website at intechopen.com.

Meet the editor

Mousumi Sen has a doctorate degree from the Indian Institute of Technology, Delhi, India. She is currently an Assistant Professor in the Department of Applied Chemistry, Amity University. Her research focuses on the prediction of pollutant dispersion from industrial areas, development of effective and sustainable methods for the removal of inorganic and organic pollutants from polluted water, food chemistry, heavy metal detoxification, composites/nanocomposites, water research, bio-inorganic chemistry, and nano chemistry. She has published numerous peer-reviewed research articles in journals of high repute as well as edited book chapters, authored a book, edited a book, and conference proceeding papers in her credit.

Contents

Preface

This book on *Nanotechnology and the Environment* includes an introduction to nanotechnology contributed to by world-renowned specialists in the field of nanomaterials, renewable energy, and environmental sciences. This book is a collection of advanced areas of research on all aspects of nanotechnologies. Nanoscience, nanotechnology, and environmental research are truly multidisciplinary and international efforts, which cover a wide range of scientific disciplines such as physics, chemistry, biology, biomedical, medicine, materials sciences etc. Nanomaterials that usually range in size from 1 to 100 nm in at least one dimension. The properties of nanomaterials are different from those of macro particles because the chemical and physical properties are very dissimilar when the dimensions are at the nano meter range and this reflects in production of very unique physiochemical properties like shape, size, dispersion, surface functionalization or de-functionalization etc. The core scientific principles and applications of this interdisciplinary field bring together chemists, physicists, materials scientists, and engineers to meet the potential future challenges such as renewable energies for sustainable development, new technologies for preparation of advanced materials, and environmental protection. This also shows a huge amount of potential application in solar cells, fuel cells, secondary batteries, capacitors, MOFs etc.

Application of nanotechnology in solar cells reflects awareness about the newly developed nanomaterials. The new architectures for solar cells focus on nanotechnology by making transparent solar cells as compared to the conventional materials. The application of nanotechnology-based photo catalysts in solar cells i.e., uses of photo nano catalyst have been appreciated as very interesting and of course environmentally sustainable i.e., self-cleaning process using nano photo catalyst.

The potential of nanotechnology in agricultural sciences using the revolutionary application of nano fertilizers have proved to be economically feasible and socially sustainable technology as compared to conventional materials. The application of nanotechnology-based nano fertilizers has potential to reduce nutrient loss due to leaching, emissions, and long-term incorporation by soil microorganisms i.e., self-cleaning process using nano fertilizers.

Modern trends in uses of different wastes to produce nanoparticles and its environmental applications reflects a healthy awareness about the newly developed nano materials. The sequences of the uses of the new architectures for synthesis of different nanoparticles such as nano-cellulose, metals and metal oxide nanoparticles, carbon nanoparticles and nano-fibers using different methods on agricultural wastes and industrial wastes have been developed. Also discussed is the self-cleaning process using nano materials from wastes by using food wastes in nanoparticle synthesis as compared to the conventional materials.

The designing and the sophisticated fabrication of nanomaterials have been discussed along with their potential energy and environmental applications. This book is a significant contribution towards the development of the knowledge for all advanced undergraduate, graduate level students, researchers, and professional engineers leading in the fields of nanotechnology, nano chemistry, macromolecular science and those who have interest in energy and environmental science.

Mousumi Sen
Amity University,
UP, India

Section 1

Nanomaterials: Synthesis and Applications

Chapter 1

Modern Trends in Uses of Different Wastes to Produce Nanoparticles and Their Environmental Applications

Salah Abdelbary and Hadeer Abdelfattah

Abstract

Wastes are produced at large amounts all over the world. These wastes cause a variety of problems to the ecosystem, plants, animals, and humans. In this chapter, we discuss the wastes, types of wastes, sources of wastes, and problems related to wastes, especially health-related problems. Then we discuss agricultural wastes and how we can synthesize different nanoparticles from them. Also, we discuss industrial wastes and different nanoparticles synthesized from them. Additionally, we discuss fruit wastes and production of different nanoparticles and also food wastes and their uses in nanoparticle syntheses. Also, we can use other wastes to produce nanoparticles. In applications section, we discuss the use of different nanoparticles produced in agriculture, removal of heavy metals and pollutants from environment, industry and finally medical applications. We will finish our chapter with the topic of healthy and safe synthesis of nanoparticles produced by different wastes and then conclusion.

Keywords: wastes, nano-cellulose, metal oxide nanoparticles, nano-carbon and environmental pollutants

1. Introduction

Wastes are unwanted, unused or useless, and disposed of after primary use. On the other hand, a product in the product can be a collective product with small measures. They are disposed of or are intended for demolition or are required to be demolished in accordance with the provisions of national law [1]. Waste materials can be turned into products or resources by innovations that increase the value of waste products above zero. Waste can also be created during raw material extraction or recycling into intermediate and final products [2]. Also, the waste is solid or gaseous, and it is chemically toxic or harmful, esthetically offensive, or radioactive. Some waste involves only temporary repression, while others can be isolated indefinitely [3]. The list of potential health problems that are important in the context of hazardous waste exposure was presented by the Agency for Toxic Substances and Diseases and includes birth faults and reproductive complaints, cancers, immune illnesses, renal and liver dysfunction, respiratory diseases, and neurotoxic disorders [4].

Green synthesis of nanoparticles has recently aroused great interest due to its advantages such as being economic, simplicity, environmental friendliness, biosynthesis, and widespread use in conventional chemical and physical methods [5]. Nanomaterial production originates from a variability of wastes including crop remains, industrial wastes, and food wastes. To this end, a variety of treatment methods have been developed and implemented to convert waste into useful nanotubes by chemical and thermal action, pumping, gas condensation dust, reduction of sodium borohydride, and thermal method of the solvent [6].

The application of nanotechnology in various fields such as health and medicine, electronics, energy, and the environment are wide. In the field of water purification, nanotechnology offers the opportunity to effectively remove dirt and bacteria. Adsorption has proven to be the best process for water purification technology due to its key advantages [7]. The use of nanotubes in drug delivery, protein delivery, and cancer peptide delivery has been explained. Different types of nanoparticles in cancer treatment are provided, such as carbon dioxide nanoparticles and wire nano-shells [8].

In this chapter, we discuss the sources and types of different wastes, their problems on the environment, their use as a source of synthesis of nanoparticles, and then the application of these produced nanoparticles in different applications especially in the environment.

2. Wastes

2.1 Sources and types of wastes

2.1.1 Agricultural wastes

Agricultural wastes can be defined as residues from the cultivation and processing of raw agricultural products such as fruits, vegetables, meat, poultry, dairy products, and crops. Agricultural waste can be solid, liquid, or lubricant depending on the nature of the agricultural activity. In addition, agricultural wastes play an important part of global agricultural productivity [9]. Agricultural waste, which includes both organic (organic) and nonorganic wastes, is a general term used to describe farm-generated waste through various agricultural activities. These activities may include, but are not limited to, milk, horticulture, seed production, animal husbandry, garden, nursery, and even forestry. Wastes from agriculture and the food industry make up a significant portion of global agricultural productivity. It is estimated that this waste may account for more than 30% of global agricultural productivity [10]. Agricultural waste today is very challenging, and many agricultural wastes are present in our environment every day. The latest trend in biofuel production from agricultural waste is in-depth research. Various processes such as chemical heat, gas emissions, liquid emissions, combustion, combustion, and rapid pyrolysis processes can be studied to obtain biofuels from agricultural wastes such as corn, straw, wheat, and rice straw [11].

2.1.2 Industrial wastes

Industrial wastes present at a huge amounts and cause a lot of pollution. There is a global consensus to reduce such waste to reduce biological burden [12]. Contribution involves industrial waste recovery, especially from mining and metallurgical enterprises. Waste is processed in the form of hydraulic loads that can be disposed of in closed underground mines [13]. Industrial waste must be compatible (treatable) with sewage. Industrial waste must be limited and proportional to the flow and burden of sewage pollution. Industrial waste shall not be toxic or harmful to the operating purpose of the treated plant material. Industrial waste should not contain harmful substances to service personnel or those from the environment near the septic tank [14]. Industrial waste is either directly connected to streams or other natural water bodies or dumped into sewers. In this way, this waste, in one way or another affects the normal life of the stream or the normal functioning of sewers and treated plants. Water can discharge a certain amount of waste before it gets dirty, and municipal wastewater treatment plants can be designed to treat all kinds of industrial waste [15]. Typically, industrial waste can be divided into two categories, hazardous and nonhazardous. Nonhazardous industrial wastes do not cause environmental and health hazards and are produced from cardboard, plastic, iron, glass, stone, and organic waste. In contrast, hazardous wastes are industrial waste that can be harmful to public health or the environment, such as flammable, biodegradable, and hazardous materials [16]. Industrial waste is classified as wastewater, solid waste, or air leaks. There is some overlap in the physical properties of the substances present in these three categories, as wastewater can contain suspended solids and suspended liquids and precipitation of solid waste can include gas, liquid, and some liquids. Particles and air exposures may consist of a fluid that emits air fluid and a substance known as particle emission [17]. Industrial waste, which has a significant concentration of nonrecyclable or recyclable metals, is usually a good candidate for landfill, which is the dumping of waste into the ground area [18].

2.2 Environmental problems of wastes

2.2.1 Agricultural waste problems

Air pollution as a result of agriculture Very low but emissions from agricultural machinery and farm wastes is a common in many developing countries. Agriculture is a major source of water pollution and land resources. In view of the large water pollution caused by agriculture, special emphasis is placed. Leaking agricultural commodities can also be fatal to human's health problems. For example, carrying the pathogen can increase significantly by leaked and stagnant water bodies. When these sources used to meet drinking water needs, water infections can occur, especially in rural areas [19]. Pesticides, fertilizers, and agricultural wastes can cause severe water and soil pollution in the region. In recent years, it has also been clear that agriculture has been a major source of air pollution, with consequences that are long-term and universal [20]. Recognize that some nitrate may exist in nature with low water concentration, and any form of agriculture is likely to raise the level of nitrate [21].

2.2.2 Industrial waste problems

Industrial pollution continues to be a major factor in worsening the environment around us, the water we use, the air we breathe, and the land we live in. The growing power of industrialization has not only consumed large agricultural land but at the same time has caused environmental degradation as well as land. Water from various industries finds its place in agriculture [22]. Waste released by industries such as sugarcane, sugarcane and resin, textile, viscose, latex, and oxalic acid have been evaluated and proven useful in agriculture. Other wastes such as sewage and sediment, fly ash, flowerpots, mud and biogas, and biowaste also have proved to be useful for increasing plant production and fertilizer savings [23]. In the course of waste production, solid and liquid industrial wastes were created. There are many elements that can be valuable components for agriculture and fertilizer and produced from industrial wastes. These wastes accumulate in significant quantities due to the high chemical content [24]. Human activities such as extraction and emission, burning of fossil fuels and fossils, as well as the use of organic and other chemicals and radium in agriculture and industry pose a risk to the environment and the general population. Awareness of these risks due to industrial waste has emerged through many cases of severe environmental impacts many years after disposal [25].

3. Synthesis of nanoparticles using different wastes

Agricultural wastes consist of both natural and nonnatural wastes such as bananas or oranges, wheat, straw, cotton or corn, coconut or almonds, silk, corn, oats, coconut oil, grapes, and empty grapes that can also be successfully applied to obtain nanoparticles [26]. On the other hand, industrial wastes have a wider variety and additional concentrated shape of hazardous materials needing special technologies and handling procedures for treatment of produced nano-materials [27].

3.1 Nano-cellulose

Nano-sized cellulose materials are currently made from agricultural wastes and involved in the durable materials industry. The main groups of nano-celluloses (NC) are two (1) nano-fibrillated cellulose (NFC) and (2) cellular nano-crystals (CNC). They are often referred to as second-generation renewable resources for oil products. Further attention has been paid to these materials due to their low density and high mechanical properties, renewability, and biogas characteristics [28]. Extraction of nano-cellulose from agricultural wastes is a promising substitute for waste treatment, and greater use of nano-cellulose in biological sciences is expected in the future [29]. Nano-cellulose has become an important topic for many research areas because of its renewable availability of biocides and many good properties [30]. In recent years, research on nanoparticles has led to many applications and focuses on the latest developments in the value of lingo-cellulosic biomass obtained from different agro-industrial crops as a source of NC, which include (i) the structure of lingo-cellulosic biomass and its effects on nano-cellulose properties and (ii) prebiological treatment and nano-cellulose extraction procedures [31]. Also, Banana bark is a type of waste that is a promising material for the production of nano-zulose. It characterizes nano-cellulose from the inner and outer layers of the pseudo-banana tree as a preliminary research strategy for designing mutant packaging material from banana nano-cellulose [27]. From industrial wastes, different alternative pathways for the production of nano-cellulose crystals have been studied due to this common acid. The hydrocarbon-producing process leads to many environmental issues such as wastewater generation and water use or access to products containing sulfur [32].

3.2 Metals and metal oxide nanoparticles

Metals and metal oxide nanoparticles can be synthesized and improved in its properties using different wastes. Fe_3O_4 nano-composites are synthesized using papaya leaves as lingo-cellulosic agricultural wastes using a simple thermal decomposition method [33]. It has been found that the development of NPs from different plant systems is cost-effective, environmentally friendly, easy, and exciting way to other procedures. The roots of the plants have preserved several minerals and food reserves. They also contain phenols, alkaloids, flavonoids, terpenoids, proteins, enzymes, carbohydrates, and other organic compounds. These metabolites play a key role in reducing metal ions in the desired NPs and also act as closing and stabilizing agents [34]. Based on the nontoxic nature of SiO_2NP from bamboo leaves, the researchers successfully synthesized 13.8 nm SiO_2NP as a source of silica, which they considered as a potential alternative for drug delivery and other medical applications [35]. The addition of banana peel extract to an alkaline solution of tetraethyl orthosilicate in ethanol, followed by calcination of the precipitate, resulted in 20 nm SiO_2NPs [36]. Sugarcane baggage was used for size control (TiO_2NPs). TiO_2 sol obtained from titanium tetra isopropoxide at pH = 4 was calcined at 200°C for 5 h, resulting in a TiO_2 powder gel [37]. Also, Mn_3O_4 nanoparticles can be synthesizes using banana peel extract which play a dual role in reducing $KMnO_4$ to Mn_3O_4 formation and preventing agglomeration of nanoparticles during preparation [38]. Tea wastes are used to synthesize hydrated aluminum nanoparticles. This porous nanomaterial is synthesized by co-precipitation between aluminum sulfate and NaOH in the presence of tea waste and anionic polyacrylamide. Maintained porous aluminum is used as an anionic exchange of fluoride with sulfate ions to neutralize drinking water [39]. Scientists have developed a unique method for the synthesis of high quality GO and reduced graphene oxide (rGO) sheets of various naturally available green wastes and carbon wastes, including animal wastes, vegetable wastes (leaves, wood, and fruit waste) and semi-industrial wastes such as newspaper [40]. Green synthesis of silver nanoparticles (AgNP), using agricultural waste, is low-cost and safe for nature and is environmentally friendly. Coconut shell extract (*Cocos nucifera*) is used to synthesize Ag NPs [41]. The SnO_2 and Ag nanoparticles were produced with a solution of nitric acid from a raw material obtained by leaching printed circuit boards. First, the tin oxide is squeezed from nitric acid solution by three different techniques: (1) normal heating, (2) microwave heating, and (3) ultrasonic treatment. Second, this precursor is transformed into tin oxide nanoparticles by furnace heat treatment. Third, hydrochloric acid is added to the nitric acid solution to cause precipitation of silver chloride. Fourth, silver chlorine is reduced to silver nanoparticles in ammonia solution, using glucose as a reducing and closing agent. The reduction reaction was performed by (i) normal warming, (II) microwave scavenging, and (III) ultrasound therapy [42].

3.3 Carbon nanoparticles

Nanocarbons were synthesized in different ways, Such as synthesis of carbon nanotubes from waste (disposable container made of polyester) using a reactor and heating system. In the reactor used, because of the high pressure and temperature above 700°C used along with the appropriate catalysts for different periods, all the materials lose their macroscopes and disperse into nanoparticles. [43]. Also, carbon nanotubes were obtained by monitoring pyrolysis of acrylic fiber residues under a layer of charcoal using physical activity in a high-temperature oven [44]. Carbon nanotubes can be released from nanoparticles into the environment at the end of their life, or whether they remain embedded in the matrix. Carbon nanotubes from poly lactic films and poly lactic acids were studied for the scenario of biodegradation and nano-composite [45]. Additionally, carbon nanoparticles synthesized by laser pyrolysis of hydrocarbons in a flow reactor have been studied as a function of laser energy [46]. On the other hand, waste plastic caused serious environmental problems. In this case, nitrogen-doped porous carbon nano-sheets (N-PCN) were prepared using magnesium hydroxide sheets [Mg (OH) 2], which are modified by Zn and Co bimetallic zeolite imidazolate frame nanoparticles such as templates and polystyrene (PS) as a carbon precursor [47]. Also, nano-channeled ultra-fine carbon tubes (NCUFCTs) and polygonal carbon nanotubes (MWCNTs) were prepared from polyethylene terephthalate (PET) waste by the spinning cathode technique. The manufacture of carbon black from anode covers ultra-fine and nano-sized solid carbon spheres (SCS) by means of an average diameter of 221 and 100 nm individually, shaped in the low-temperature area of the anode, where the temperature is around 1700°C [48]. Carbon-bound nanofibers (CNFs) are obtained by the decomposition of methane onto Ni nanoparticles supported by grooved SiC nanowires. In beam CNFs, several CNFs grow in parallel and form a packet of CNFs [49]. Tea wastes are rich in carbon, nitrogen, and potassium, but poor in phosphorus, which means that they can also be used to reduce metal oxides once they are carbonated and form carbon nanoparticles [50]. Also, thermoplastic polymers (such as polypropylene, polyethylene, polyvinyl chloride, polystyrene, etc.) are the main components of municipal solid waste. Millions of tons of plastic waste are dumped every year, most of which is incinerated or dumped. Alternatively, various researchers have proposed methods using this waste as feed to produce value-added products such as fuels, carbon nanotubes, and porous carbon emissions [51].

3.4 Silica and graphene nanoparticles

The production of silica nanoparticles by conventional processes is complex and takes place at very high temperatures. Silica nanoparticles of different sizes are obtained from plastic waste, disposable boxes, and water bottles by a simple method of carbothermal reduction [52]. Also, researchers developed an alternative use of some agricultural waste as potential sources of silicon that can be used for PV cells. This study examines the use of cassava periderm, corn stalk, and cob as new sources of silicon nanoparticles. Agro-based silicon nanoparticles are prepared by the modified sol-gel method and then reduced using magnesium to synthesize silicon nanoparticles [53]. A nano-composite material found on Ag nanoparticles and graphene oxide was considered for its electrical, optical, and physical properties. According to electron and atomic force microscopy data, the size of the nanoparticles obtained varies mostly from 60 to 100 nm. The permeability and electrical resistance of this material indicate higher optical transparency and electrical conductivity than in virgin graphene oxide [54]. Agricultural wastes as rice straw, rice husk, and leaves of bamboo delivered a simple method to silicon production. Several agricultural wastes are generated and disposed of indiscriminately in the environment and thus pose environmental challenges. This study examines the use of cash periderm as a new source of silicon nanoparticles. The cassava was treated with acid before and after fermentation to obtain silicone residues for the gelation process for the production of silicon nanoparticles [55]. The synthesis of silica from nanoparticles from rice husk, sugarcane, and coffee nut has been reported using vermicompost with Anelides (*Eisenia foetida*). The product (humus) is calcined and extracted to recover the crystalline nanoparticles [56].

4. Environmental applications of produced nanoparticles using different wastes

Pollution is one of the biggest problems in the world, which poses a lot of risks to humans, animals, plants, and ecosystems [57]. Nano-cellulose has a diameter usually <10 nm, which gives it many unique properties. Among many others, these properties include high mechanical strength, large area, and low visual light scattering [58]. The four main groups of cellulosic nanoparticles and their easy surface modification provide a huge variety of new materials, composites, films, and gels with captivating and controllable properties to solve environmental problems and challenges [59]. Applications and properties of cellulosic nanoparticles such as adsorbent, photocatalyst, flocculant, and membranes have been reviewed in particular [60]. Attractive properties facilitate the use of nano-cellulose aerogels in various environmental and engineering applications such as water purification, filtration, flame retardation, and oil extraction [61]. The nano-cellulose has become a sustainable and successful nanomaterial with its unique structure and features such as high specific modulus, excellent stability in most solvents, low toxicity, and natural diversity. Eco-friendly environment, low cost, convenience, and simple synthesis techniques make nano-lotus a promising candidate for the production of green renewable energy storage [62]. Environmental challenges that can be addressed by the use of metal oxide nanoparticles that are produced by different wastes include removal of toxic chemicals such as different heavy metals from industrial wastewater and wastewater, catalysts for organic reactions to produce essential organic material, reproducible genes for environmental restoration, volatile organic compounds and detectors, and biological/chemical signals [63]. Catalysts of metal oxide nanoparticles include reaction-base, selective oxidation, complete oxidation, depolation, biosynthesis, green chemistry, and photocatalysis. Iron oxide nanoparticle catalysts are important components in the refining and petrochemical processes. These catalysts are also important for improving environmental quality [64]. On the other hand, graphene and carbon tube nanoparticles offer a variety of advanced applications in the field of energy storage, biological applications, and electrolytes due to their mechanical, electrical, electrical, and thermal properties [65]. Nanocarbon-TiO_2 composites were prepared by the liquid phase deposition method to apply as photocatalytic for the degradation of heavy metals, diphenhydramine, and dyes [66].

5. Conclusion

This chapter concluded that, millions of tonnes of different wastes are produced annually without any benefits from it. Researchers have used these wastes in synthesis of different nanoparticles such as nano-cellulose, metals, and metal oxide nanoparticles, carbon nanoparticles and nano-fibers by different methods. These NPs are used to solve environmental problems, especially pollution, for which they are used as protective agents and adsorbents. In future prospective, researchers must use these NPs in a wide range of applications as ecofriendly and low-cost products.

Acknowledgements

Firstly, great thanks and appreciation to the staff of Al-Azhar University, Faculty of Science for their support and encouragement. Finally, I dedicate this work to my father's soul and thank my mother, brother, sisters, and everyone in my family for their continual guidance.

Conflict of interest

The authors declare no conflict of interest.

Author details

Salah Abdelbary[1]* and Hadeer Abdelfattah[2]

1 Botany and Microbiology Department, Faculty of Science, Al-Azhar University, Nasr City, Egypt

2 Drug Radiation Research Department, Atomic Energy Authority, Nasr City, Egypt

*Address all correspondence to: salah.micro87@gmail.com

References

[1] Basel Convention. Archivedcopy (PDF). 1989. Archived (PDF) from the original on 2017-05-16 [Retrieved: 25 May 2017]

[2] Glossary of Environment Statistics. UNSD; 1997. Archived (2013) at the Wayback Machine

[3] Nace RL. Problems of underground storage of wastes. Journal of Research of the U.S. Geological Survey. 1973;**1**(6):719-723

[4] Zejda JE. Health effects of hazardous wastes. Central European Journal of Public Health. 1998;**6**(2):140-143

[5] Adelere IA, Lateef A. A novel approach to the green synthesis of metallic nanoparticles: The use of agro-wastes, enzymes, and pigments. Nanotechnology Reviews. 2016;**5**(6):567-587. DOI: 10.1515/ntrev-2016-0024

[6] Samaddar P, Ok YS, Kim KH, Kwon EE, Tsang DCW. Synthesis of nanomaterials from various wastes and their new age applications. Journal of Cleaner Production. 2018;**197**: 1190-1209. DOI: 10.1016/j.jclepro.2018.06.262

[7] Ali HR. Applications of bio-waste materials as green synthesis of nanoparticles and water purification. Advances in Materials. 2017;**6**(5):85. DOI: 10.11648/j.am.20170605.16

[8] Pavlovic M, Mayfield J, Balint B. Nanotechnology and its application in medicine. In: Handbook of Medical and Healthcare Technologies. Annals of Medical and Health Sciences Research; 2013. pp. 181-205. DOI: 10.1007/978-1-4614-8495-0_7

[9] Foster CN. Agricultural wastes: Characteristics, types and management. In: Agricultural Wastes: Characteristics, Types and Management. 2015. pp. 1-15

[10] Ashworth GS, Azevedo P. Agricultural wastes. In: Agricultural Wastes. Agriculture Issues and Policies Series; 2009. DOI: 10.2175/106143099x133767

[11] Swain PK. Utilization of agriculture waste products for production of bio-fuels: A novel study. Materials Today: Proceedings. 2017;**4**(11):11959-11967. DOI: 10.1016/j.matpr.2017.09.117

[12] CA. A treatise on sub- and supercritical fluids: Versatile domains and applications. In: Reference Module in Food Science. Comprehensive Review in Food Science and Food Safety; 2019. DOI: 10.1016/b978-0-08-100596-5.22951-x

[13] Dvořáček J, Vodzinský V, Domaracká L. Industrial wastes and economics of their utilization. Meta. 2006;**45**(2):141-143

[14] Nemerow NL. Industrial collaborative solutions. In: Environmental Solutions. 2005. pp. 249-295. DOI: 10.1016/B978-012088441-4/50013-7

[15] Muralikrishna IV, Manickam V. Industrial wastewater treatment technologies, recycling, and reuse. In: Environmental Management. 2017. pp. 295-336. DOI: 10.1016/b978-0-12-811989-1.00013-0

[16] Millati R, Cahyono RB, Ariyanto T, Azzahrani IN, Putri RU, Taherzadeh MJ. Agricultural, industrial, municipal, and forest wastes. In: Sustainable Resource Recovery and Zero Waste Approaches. 2019. pp. 1-22. DOI: 10.1016/b978-0-444-64200-4.00001-3

[17] Ramachandra Rao S. Waste Characterization. 2006. pp. 13-34. DOI: 10.1016/s0713-2743(06)80087-5

[18] Artiola JF. Industrial waste and municipal solid waste treatment

and disposal. In: Environmental and Pollution Science. International Journal of Environmental Research and Public Health; 2019. pp. 377-391. DOI: 10.1016/b978-0-12-814719-1.00021-5

[19] Nagendran R. Agricultural waste and pollution. In: Waste. Elsevier; 2011. pp. 341-355. DOI: 10.1016/B978-0-12-381475-3.10024-5

[20] Pretty JN, Conway GR. Agriculture as a global polluter. Agriculture. 1989;**11**:1-16

[21] OECD: water pollution by fertilizers and animal wastes. World Farmers' Times; 1987. p. 2

[22] Saranraj P, Stella D. Impact of sugar mill effluent to environment and bioremediation: A review. World Applied Sciences Journal. 2014;**30**(3):299-316. DOI: 10.5829/idosi.wasj.2014.30.03.1656

[23] Dikshit PR, Khatik SK. Contribution and potential of industrial wastes and sewage sludge for increasing crop production. Journal of Industrial Pollution Control. 2000;**16**(1):81-93

[24] Wolski T, Glinski J. Utilization of environment-polluting industrial wastes for agriculture and the fertilizer industry. Studies in Environmental Science. 1986;**29**(C):599-607. DOI: 10.1016/S0166-1116(08)70965-8

[25] Lal R. Pollution: Industrial waste. In: Encyclopedia of Soil Science. 3rd ed. Taylor and Francis Group; 2017. pp. 1762-1764. DOI: 10.1081/e-ess3-120006661

[26] Zamani A, Poursattar Marjani A, Abdollahpour N. Synthesis of high surface area boehmite and alumina by using walnut shell as template. International Journal of Nano and Biomaterials. 2019;**8**:1-14

[27] Chandrappa R, Das DB. Wastes from industrial and commercial activities. Environmental Science and Engineering (Subseries: Environmental Science). 2012;(9783642286803):217-247. DOI: 10.1007/978-3-642-28681-0_9

[28] Rajinipriya M, Nagalakshmaiah M, Robert M, Elkoun S. Importance of agricultural and industrial waste in the field of nanocellulose and recent industrial developments of wood based nanocellulose: A review. ACS Sustainable Chemistry & Engineering. 2018;**6**(3):2807-2828. DOI: 10.1021/acssuschemeng.7b03437

[29] Sangeetha J, Thangadurai D, Hospet R, Purushotham P, Manowade KR, Mujeeb MA, et al. Production of bionanomaterials from agricultural wastes. In: Nanotechnology: An Agricultural Paradigm. Nanomaterial and Nanostructures; 2017. pp. 33-58. DOI: 10.1007/978-981-10-4573-8_3

[30] García A, Gandini A, Labidi J, Belgacem N, Bras J. Industrial and crop wastes: A new source for nanocellulose biorefinery. Industrial Crops and Products. 2016;**93**:26-38. DOI: 10.1016/j.indcrop.2016.06.004

[31] Pires JRA, Souza VGL, Fernando AL. Ecofriendly strategies for the production of nanocellulose from agro-industrial wastes. In: European Biomass Conference and Exhibition Proceedings. 2019. pp. 1781-1784

[32] Satyamurthy P, Vigneshwaran N. A novel process for synthesis of spherical nanocellulose by controlled hydrolysis of microcrystalline cellulose using anaerobic microbial consortium. Enzyme and Microbial Technology. 2013;**52**:20-25

[33] Ahmed M, Ahmaruzzaman M. Fabrication and characterization of novel lignocellulosic biomass tailored Fe_3O_4 nanocomposites: Influence of annealing temperature and chlorazol black sequestration. RSC Advances. 2015;**5**:107466107473

[34] Bachheti RK, Godebo Y, Bachheti A, Yassin MO, Husen A. Root-based fabrication of metal/metal-oxide nanomaterials and their various applications. In: Nanomaterials for Agriculture and Forestry Applications. 2020. pp. 135-166. DOI: 10.1016/b978-0-12-817852-2.00006-8

[35] Kauldhar B, Yadav S. Turning waste to wealth: A direct process for recovery of nano-silica and lignin from paddy straw agro-waste. Journal of Cleaner Production. 2018;**194**:158-166

[36] Ali SM. Fabrication of a nanocomposite from an agricultural waste and its application as a biosorbent for organic pollutants. Journal of Environmental Science and Technology. 2018;**15**:1169-1178

[37] Xue H, Chen Y, Liu X, Qian Q, Luo Y, Cui M, et al. Visible light-assisted efficient degradation of dye pollutants with biomass-supported TiO_2 hybrids. Materials Science & Engineering, C: Materials for Biological Applications. 2018;**82**:197-203

[38] Yan D, Zhang H, Chen L, Zhu G, Wang Z, Xu H. Supercapacitive properties of MnO_3 nanoparticles biosynthesized from banana peel extract. RSC Advances. 2014;**4**:23649-23652

[39] Cai H, Chen G, Peng C, Xu L, Zhu X, Zhang Z. Enhanced removal of fluoride by tea waste supported hydrous aluminium oxide nanoparticles: Anionic polyacrylamide mediated aluminium assembly and adsorption mechanism. RSC Advances. 2015;**5**:29266-29275

[40] Rajesh K, Rajesh KS, Dinesh PS. Natural and waste hydrocarbon precursors for the synthesis of carbon based nanomaterials: Graphene and CNTs. Renewable and Sustainable Energy Reviews. 2016;**58**:976-1006

[41] Sinsinwar S, Sarkar M, Suriya K, Nithy P, Vadivel V. Use of agricultural waste (coconut shell) for the synthesis of silver nanoparticles and evaluation of their antibacterial activity against selected human pathogens. J. Micpath. 2018;**124**:30-37

[42] Cerchier P, Dabalà M, Brunelli K. Synthesis of SnO_2 and Ag nanoparticles from electronic wastes with the assistance of ultrasound and microwaves. JOM. 2017;**69**(9):1583-1588. DOI: 10.1007/s11837-017-2464-x

[43] Kaviani D. Synthesis of carbon nanoparticles from polystyrene wastes. International Journal of Sciences. 2015;**1**(8):53-57. DOI: 10.18483/ijsci.797

[44] Baheti V, Naeem S, Militky J, Okrasa M, Tomkova B. Optimized preparation of activated carbon nanoparticles from acrylic fibrous wastes. Fibers and Polymers. 2015;**16**(10):2193-2201. DOI: 10.1007/s12221-015-5364-0

[45] Kotsilkov S, Ivanov E, Vitanov NK. Release of graphene and carbon nanotubes from biodegradable poly(lactic acid) films during degradation and combustion: Risk associated with the end-of-life of nanocomposite food packaging materials. Materials. 2018;**11**(12). DOI: 10.3390/ma11122346

[46] Galvez A, Herlin-Boime N, Reynaud C, Clinard C, Rouzaud JN. Carbon nanoparticles from laser pyrolysis. Carbon. 2002;**40**(15):2775-2789. DOI: 10.1016/S0008-6223(02)00195-1

[47] Wang G, Liu L, Zhang L, Fu X, Liu M, Zhang Y, et al. Porous carbon nanosheets prepared from plastic wastes for supercapacitors. Journal of Electronic Materials. 2018;**47**(10):5816-5824

[48] Joseph Berkmans A, Jagannatham M, Priyanka S, Haridoss P.

Synthesis of branched, nano channeled, ultrafine and nano carbon tubes from PET wastes using the arc discharge method. Waste Management. 2014;**34**(11):2139-2145. DOI: 10.1016/j.wasman.2014.07.004

[49] Guo X, Guo X, Zhi G, Wang Y, Jin G. Bundle-like carbon nanofibers grown from methane decomposition. Carbon. 2012;**50**(1):321-322. DOI: 10.1016/j.carbon.2011.07.046

[50] Güler Ö, Boyrazlı M, Başgöz Ö, Bostancı B. The synthesis of carbon nanostructures from tea plant wastes. Canadian Metallurgical Quarterly. 2017;**56**(3):349-359. DOI: 10.1080/00084433.2017.1345467

[51] Bazargan A, Hui CW, McKay G. Porous carbons from plastic waste. Advances in Polymer Science. 2013;**266**:01-26. DOI: 10.1007/12_2013_253

[52] Meng S, Wang DH, Jin GQ, Wang YY, Guo XY. Preparation of SiC nanoparticles from plastic wastes. Materials Letters. 2010;**64**(24):2731-2734. DOI: 10.1016/j.matlet.2010.09.007

[53] Adebisi JA, Agunsoye JO, Ahmed II, Bello SA, Haris M, Ramakokovhu MM, et al. Production of silicon nanoparticles from selected agricultural wastes. Materials Today: Proceedings. 2020. DOI: 10.1016/j.matpr.2020.03.658

[54] Neustroev EP, Kurkina II, Mamaeva SN, Nogovitsyna MV. Synthesis, characterisation and applications of nanocomposites based on silver nanoparticles and graphen oxide. Journal of Structural Chemistry. 2018;**59**(4):847-852. DOI: 10.1134/S0022476618040145

[55] Agunsoye JO, Adebisi JA, Bello SA, Haris M, Agboola JB, Hassan SB. Synthesis of silicon nanoparticles from cassava periderm by reduction method. In: Materials Science and Technology 2018, MS and T 2018. 2019. pp. 701-709. DOI: 10.7449/2018/MST_2018_701_709

[56] Espíndola-Gonzalez A, Martínez-Hernández AL, Angeles-Chávez C, Castaño VM, Velasco-Santos C. Novel crystalline SiO_2 nanoparticles via annelids bioprocessing of agro-industrial wastes. Nanoscale Research Letters. 2010;**5**(9):1408-1417. DOI: 10.1007/s11671-010-9654-6

[57] Abdelbary S, Elgamal MS, Farrag A. Trends in heavy metals tolerance and uptake by *Pseudomonas aeruginosa*. In: Sriramulu D, editor. *Pseudomonas aeruginosa*—An Armory Within. IntechOpen; 2019. DOI: 10.5772/intechopen.85875

[58] Wei H, Rodriguez K, Renneckar S, Vikesland PJ. Environmental science and engineering applications of nanocellulose-based nanocomposites. Environmental Science: Nano. 2014;**1**(4):302-316. DOI: 10.1039/c4en00059e

[59] Soriano ML, Ruiz-Palomero C. Nanocellulose as promising material for environmental applications. Nanotechnology in Environmental Science. 2018;**2-2**:579-598. DOI: 10.1002/9783527808854.ch18

[60] Shak KPY, Pang YL, Mah SK. Nanocellulose: Recent advances and its prospects in environmental remediation. Beilstein Journal of Nanotechnology. 2018;**9**(1):2479-2498. DOI: 10.3762/bjnano.9.232

[61] Gopakumar DA, Thomas S, Owolabi FAT, Thomas S, Nzihou A, Rizal S, et al. Nanocellulose based aerogels for varying engineering applications. In: Encyclopedia of Renewable and Sustainable Materials. 2020. pp. 155-165. DOI: 10.1016/b978-0-12-803581-8.10549-1

[62] Jose J, Thomas V, Vinod V, Abraham R, Abraham S. Nanocellulose based functional materials for supercapacitor applications. Journal of Science: Advanced Materials and Devices. 2019;**4**(3):333-340. DOI: 10.1016/j.jsamd.2019.06.003

[63] Ganachari SV, Hublikar L, Yaradoddi JS, Math SS. Metal oxide nanomaterials for environmental applications. In: Handbook of Ecomaterials. Vol. 4. 2019. pp. 2357-2368. DOI: 10.1007/978-3-319-68255-6_196

[64] Védrine JC. Heterogeneous catalysis on metal oxides. Catalysts. 2017;**7**(11). DOI: 10.3390/catal711034

[65] Saba N, Jawaid M, Fouad H, Alothman OY. Nanocarbon: Preparation, properties, and applications. In: Nanocarbon and its Composites: Preparation, Properties and Applications. ACS Publications; 2018. pp. 327-354. DOI: 10.1016/B978-0-08-102509-3.00009-2

[66] Pastrana-Martínez LM, Morales-Torres S, Papageorgiou SK, Katsaros FK, Romanos GE, Figueiredo JL, et al. Photocatalytic behaviour of nanocarbon-TiO2 composites and immobilization into hollow fibres. Applied Catalysis B: Environmental. 2013;**142-143**:101-111. DOI: 10.1016/j.apcatb.2013.04.074

Chapter 2

Application of Nanomaterials in Environmental Improvement

Ali Salman Ali

Abstract

In recent years, researchers used many scientific studies to improve modern technologies in the field of reducing the phenomenon of pollution resulting from them. In this chapter, methods to prepare nanomaterials are described, and the main properties such as mechanical, electrical, and optical properties and their relations are determined. The investigation of nanomaterials needed high technologies that depend on a range of nanomaterials from 1 to 100 nm; these are scanning electron microscopy (SEM), transmission electron microscopy (TEM), and X-ray diffractions (XRD). The applications of nanomaterials in environmental improvement are different from one another depending on the type of devices used, for example, solar cells for producing clean energy, nanotechnologies in coatings for building exterior surfaces, and sonochemical decolorization of dyes by the effect of nanocomposite.

Keywords: nanomaterials, synthesis, solar cell, sonocatalyst, water purification

1. General introduction

The term nanotechnology is the creation of functional material devices and systems through the control of matter in the range of 1–100 nm and the ability to work at the molecular level, atom by atom to create large structures with fundamentally new molecular organization. Nanotechnology is the design, fabrication, and application of nanostructures or nanomaterials and the fundamental understanding of the relationships between physical properties, or phenomena, and material dimensions. It is a new field or a new scientific domain. Nanometer (nm) is one billionth of a meter (10^{-9} m). About 10 hydrogen or fife silicon atoms are arranged in a straight line approximately representing 1 nm in length, and these materials are characterized by at least one dimension in the nanometer range.

Nanomaterials are of interest because at this scale unique optical, magnetic, electrical, and other properties emerge. These emergent properties have great potential applications in electronics, medicine, and other fields. Nanomaterials are classified into nanostructured and nanophase/nanoparticle materials. The former refer to condensed bulk materials that are made of grains with grain sizes in the nanometer size range, while the latter are usually the dispersive nanoparticles [1]. According to this definition, a nanoparticle is considered to have zero dimensions (the dimensions' length is less than 100 nm). For example, wires, rods, and nanofibers are objects with one dimension, while thin films, plates, multilayers, and network nanostructures express two dimensions. And more clearly, a sphere or cluster of nanophase materials of zero dimension is represented as a point-like particle that is determined by three dimensions of nanomaterials, as demonstrated in **Figure 1** [2]. There are several important applications of nanomaterials such as aviation and space, chemical industry, optics, solar hydrogen, fuel cell, batteries, sensors, power generation, aeronautic industry, building/construction industry, automotive engineering, consumer electronics, thermoelectric devices, pharmaceuticals, and cosmetic industry [3]. One of the most pressing challenges of our time is to find alternative energy sources which are environmentally friendly which is depending on used of nanomaterial's in different applications such as solar cell [4], paints [5] and other applications in the field of green chemistry [6].

2. Surface effects

Chemical and physical properties of a material, such as bulk or nanoscale, depend on its surface properties. But the volume of bulk materials remains unchanged when it is subdivided into an ensemble of individual nanomaterials, and the collective surface area is greatly increased [7]. **Figure 2** describes stages of surface to volume increase for bulk materials.

Melting temperature of nanomaterials depends on the number of surface atoms and the increases of surface to volume ratio (S/V) lead to decreases in particle size and melting point because of surface atoms that have a much greater effect on chemical and physical properties of nanoparticle [8]. The surface to volume ratio for a material or substance made of nanoparticles has a significant effect on the properties of the material, when materials made up of nanoparticles have a relative larger surface area and compared to the same volume of material made up of bigger particles. For example, the surface area of a sphere, $A = 4\pi r^2$, divided by its volume, $V = \frac{4}{3}\pi r^3$, produced a *3/r* as ratio between them, or in terms of diameter *d* produced the ratio (*6/d*). The ratio ($F = {}^{A}\!/_{V}$) for large thin plates with thickness *d* is

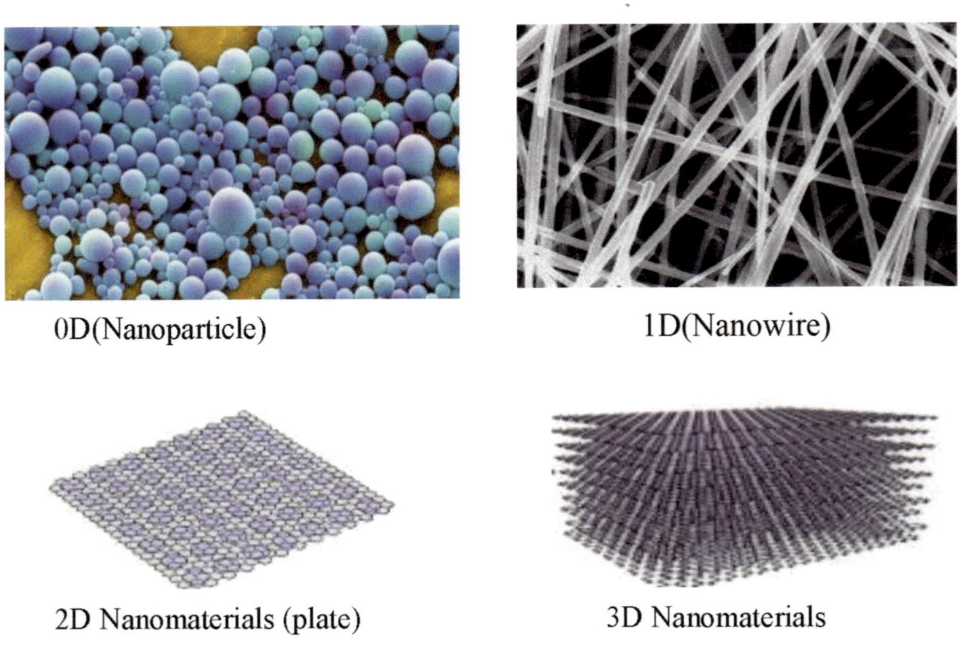

Figure 1.
Types of nanomaterials (0D, 1D, 2D, and 3D).

Figure 2.
Schematic drawing showing how surface to volume increases with decreased size.

equal to (*1/d*) and this is the same as that for long cylindrical wires. Thus, the dispersion scales *F* equals to (*1/d*) or (*1/r*) for anything having a very small range of thickness *d*. The dispersion *F* represented the fraction of atoms at the surface and it scales with surface area divided by volume of sphere scales with the square of its radius *r*, but its volume scales with r^3. The total number of atoms *N* in this sphere scales linearly with volume. The corner correction and the edge for large *N* can be negligible, leading to the $N^{-1/3}$ scaling [9]:

$$F = \frac{6n^2 - 12n + 8}{n^3} \tag{1}$$

Or nearly equal to:

$$F \approx \frac{6}{N^{1/3}} \tag{2}$$

3. Prepared nanoparticle

The two basic approaches to creating nanomaterials in a controlled and repeatable manner are the "top-down" and "bottom-up" techniques as shown in **Figure 3**, either for atoms to assemble together (break) or disassemble (dissociate) bulk solids into small pieces or to get on a few atoms from them. This is very important to use at different application fields, for example, in engineering, chemistry, physics, and even medicine. Former approaches play a very important role in modern industry and most likely in nanotechnology as well. In general, nanomaterials can be produced by different methods: mechanical, chemical, hydrothermal, sol-gel, chemical deposition in vacuum, pyrolysis, combustion, chemical co-precipitation, etc. According to these methods, particles are defined by a certain dimensional morphology and distribution can be obtained.

3.1 Top-down

In the physical methods, mechanical methods offer the least expensive ways to produce nanomaterials in bulk (break the particles into nanostructures). But

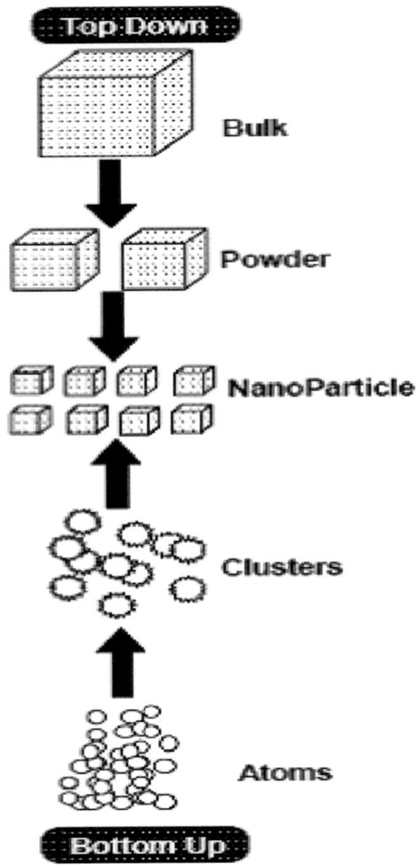

Figure 3.
The scheme to prepare nanomaterial.

chemical fabrication methods are always easy to upscale and many, such as anodizing, are widespread industrial processes [10]. Top-down approach is the process of making nanostructures that start with larger structures and break away to nanosize to form nanomaterials. To obtain nanoscale structures in this method, first, a large object that is (2–3) orders larger in one or two dimensions than the nanoscale desired is fabricated and then nanopatterning techniques are utilized to achieve smaller features. Top-down methods actually was developed firstly by and has been widely used in microelectronics industry. Methods of deposition and nanopatterning of thin films are more advanced, and this approach has been pushed further into the regime of nanofabrication [11]. Also, applying the top-down assembly process of nanocomponents over large areas is difficult and expensive.

3.2 Bottom-up

The building of nanostructures starting with small components such as atoms or molecules is called bottom-up approach. The bottom-up techniques make use of self-processes for ordering of supramolecular or solid-state architectures from the atomic to the mesoscopic scale. The methods of bottom-up include gas-phase and liquid-phase methods. For two methods, fabrication of nanomaterials was controlled when starting from the single atom or molecule. Chemical vapor deposition (CVD) and plasma arcing are called gas-phase methods, whereas liquid-phase (LP) represented by the most established method is sol-gel synthesis. Also, a new method called molecular self-assembly emerged. The areas of application for nanotechnology have different fields such as photonics, electronics, chemical sensors, biological sensors, and energy storage, and catalysis nanomaterial requires the manipulation into functional materials and devices. Self-assembly is the method important for designing and controlling the bottom-up assembly of the materials in the nanoscale range into structures of sheets, tubes, wires, nanoelectronic devices and drug delivery systems [12].

4. TiO_2/α-Al_2O_3 nanocomposite

In this section, a mechanism for preparing nanomaterials such as TiO_2, Al_2O_3, and TiO_2/α-Al_2O_3 will be explained according to the method of preparation by using sol-gel methods. The sol-gel method was developed in the 1960s mainly due to the need of new synthesis methods in the nuclear industry. The sol-gel process is defined as a gelation means that changes materials by polycondensation reactions from liquid state to gel state. If the dispersion of colloidal particles or polymers is stable in a solvent, it is called a sol, but particles can be amorphous or crystalline in the size of few nanometers. And on the other side, the gel consists of sol particles as continuous network in 3D, enclosed in a liquid phase [13]. There are several methods to prepare TiO_2 nanoparticles using different materials such as tetraisopropyl orthotitanate (TIPT), titanium tetrachloride ($TiCl_4$), ethanol (EtOH), methanol (MeOH), n-hexane, hydroxypropyl cellulose (HPC), 1,4-cyclohexanediol (CHD), triethanolamine (TEA), and TiO_2-P_{25}. In general, it is obtained on gel solution; the gel was filtered and washed subsequently by water and ethanol and then dried at room temperature to get on TiO_2 nanoparticles [14]. Nanocomposites of TiO_2 can be used at different applications in a heterogeneous catalysis, in application of photocatalyst, to produce a hydrogen and electric energy by using a solar cells, gas sensor, white pigment for a paints and cosmetic products, corrosion-protective coating, optical coating, and in electric devices varistors and etc.

On the other hand, Al_2O_3 nanoparticles are prepared by ethanol solution of aluminum nitrate Al $(NO_3)_3.9H_2O$ dissolved in pure water and then added to the solution ethanol from time to time until the color changes. The potential of hydrogen or acidic function (pH) was maintained between 2 and 3 during the synthesis. The white product was evaporated and the result was cooled to room temperature and then finally calcined at high temperature to get on nanoparticles [15]. For TiO_2/α-Al2O_3, nanocomposite was prepared by adding TTIP to isopropyl alcohol under constant stirring and at room temperature (RT = 27°C) and then dispersed of nano-alumina in TTIP solution to form white suspension. Under vigorous stirring, the white gel was formed. Then, this gel was heated at high temperature in a Teflon-lined autoclave. Finally, the collection powder of nanocomposite was yields by during the gel [16].

5. Properties of nanomaterials

The nature of all materials in bulk has different properties, which are depended on their structural properties (metals, semiconductors, and insulators), such as electrical, optical, and mechanical properties. Nanoparticles have properties that are different from small molecules; in this case, their chemistry and synthesis can be considered like complex mixtures.

The ability of the molecules to contact of nanoparticles on the surface and exchange with other molecules leads to the indicates that will be careful consideration of the chemistry of nanoparticles and how it relates to their fate in surface waters and sediments, this is a key to predicting their final fate [17]. When one of the three spatial dimensions is of a size comparable or smaller to wave length of de Broglie (λ_B) of the charge carrier of electrons and holes or the wavelength of light, the crystalline materials are destroyed by the periodic boundary conditions or change the atomic density on the surface of amorphous materials. Because of this property, a lot of the physical properties of nanomaterials are quite different from bulk materials, yielding a wide variety of new applications [18].

5.1 Mechanical properties

The mechanical properties of materials depend essentially on the nature of bonding that holds their constituent atoms and their microstructures in a variety of length scales. Mechanical deformation can be either elastic (reversible) or plastic (irreversible) [19]. Elastic materials respond to stress fields via strain fields; liquids respond via viscous strain rates; and complex fluids are often describable via frequency-dependent viscoelastic responses. Many properties of crystals, magnets, liquid crystals, superconductors, superfluids, and field theories of the early universe can be described by focusing on long length scales, assuming that the materials are close in equilibrium. On the other hand, plastic materials can be defined as irreversible deformation, and different mechanisms may be responsible: dislocation motion, vacancy motion, twinning, phase transformation, or viscous flow of amorphous materials [20]. The proportional relation between the stress and the elastic strain is given by Hooke's law, which can be written as follows:

$$\sigma \propto \varepsilon \tag{3}$$

$$\sigma = Y\varepsilon \tag{4}$$

where σ is the stress, ε is the strain, and Y is the modulus of elasticity or Young's modulus.

The size of grain for polycrystalline materials, depending on strength and hardness, is well established as Hall-Petch relationship, which indicates that the yield stress and hardness are inverse to the square root of the grain size. This strengthening at reduced grain sizes is attributed to the pile-up of dislocations at grain boundaries. However, when it comes to Nanocrystalline regime, the conventional Frank-Read dislocation sources ceases to control the deformation due to the stress to bow out a dislocation approaches the theoretical shear strength [21]. The relation between yield stress and grain size is described mathematically by:

$$\sigma_y = \sigma_o + \frac{k_y}{\sqrt{d}} \tag{5}$$

where Eq. (5) is called Hall-Petch relationship, and k_y is the strengthening coefficient, σ_o is a materials constant for the starting stress for dislocation movement, d is the grain diameter, and σ_y is the yield stress.

5.2 Electrical properties

The electrical conductivity, DC, for nanoparticle materials (or metals) is affected by the microstructure. The value of conductivity (DC) appears by grain boundary contribution, which depends on DC bias voltages but grain contribution does not depend on it. The grain boundaries in nanocrystalline materials often have significant influence on the flow of electronic current. The microstructure at scale of length is smaller or similar to the mean free path of conduction electrons, this produced a grain boundaries a main source of eight conduction electrons scattering [22]. The measurement of the electrical properties is also important because the connectivity of a composite system from SEM and TEM micrographs cannot be deduced alone. The DC electrical conductivity (σ_{dc}) of the crystal was calculated using the relation:

$$\sigma_{dc} = t/RA \tag{6}$$

where R is the measured resistance, t is the thickness of the sample, and A is the area of the face in contact with the electrode. The temperature variation of conductivity is given by using Stuke's Equation [23]:

$$\sigma_{dc} = \sigma_o \exp\left[-E/kT\right] \tag{7}$$

where σ_o is a constant depending on material, E is the activation energy, T is the absolute temperature, and k is the Boltzmann's constant.

On the other hand, the AC conductivity of the media (composites) (σ_m) is the sum of the real and imaginary conductivities, which are given by:

$$\sigma_m = \sigma_{mr} + i\sigma_{mi} \tag{8}$$

The conductivity of the more conducting component is given by:

$$\sigma_c = \sigma_{cr} + i\sigma_{ci} \tag{9}$$

For ideal conductivity where ($\sigma_{cr} >> \sigma_{ci}$) Eq. (9) read as:

$$\sigma_c = \sigma_{cr} \tag{10}$$

For the insulating component, the conductivity is given by:

$$\sigma_i = \sigma_{ir} + i\sigma_{ii} \tag{11}$$

where $\sigma_{ii} = \omega\varepsilon_0\varepsilon_{ir}$.

Eq. (11) can be approximated when ($\sigma_r < <i\sigma_{ii}$) as:

$$\sigma_i = i\omega\varepsilon_0\varepsilon_{ir} \tag{12}$$

In practice, σ_{ir} incorporates both, a usually very small, DC conductivity and the dielectric polarization loss term ($\omega\varepsilon_0\varepsilon_{ii}$). The expressions for σ_c and σ_i can be dispersive and/or temperature-dependent. [24].

5.3 Optical properties

When light incidents from one medium into another, several things are happened see **Figure 4**. Some of the light radiation may be transmitted through the medium, some will be absorbed, and some will be reflected at the interface between the two media. The total intensity (I_o) of the incident light striking a surface is equal to the sum of the absorbed (I_A), reflected (I_{oR}), and transmitted (I_T) intensities, that is

$$I_o = I_T + I_R + I_A \tag{13}$$

where T, A, R are transmissivity, absorptivity, and reflectivity, respectively. And

$$T = {}^{I_T}\!/_{I_o} \tag{14}$$

$$A = {}^{I_A}\!/_{I_o} \tag{15}$$

$$R = {}^{I_R}\!/_{I_o} \tag{16}$$

So Eq. (13) becomes:

$$1 = T + R + A \tag{17}$$

One can estimate the absorption coefficient (α) of thin films after the correction of reflectivity as:

$$\alpha = \frac{2.303}{t} A \tag{18}$$

where t is the thickness of the material.

It is very important to study α in order to define types of the electron transition, such as allowed direct, forbidden direct, allowed indirect, and forbidden indirect. The transition is allowed if $\alpha > 10^4$, when $\alpha < 10^4$ the transition is forbidden direct. From the absorption coefficient data, one can calculate the extinction coefficient (K) as [25]:

$$\lambda = \frac{4\pi}{\alpha} K \tag{19}$$

where λ is the wavelength of the incident light.

An alternative way to boost optical absorption is to use nanostructure-based devices to attain multiple band gaps based on the size of the quantum dots or quantum wells (based on quantum mechanics, the size of the dot or well determines the band gap of the material). For silicon as an example, the nanostructure results in direct band gap material, and the optical absorption is enhanced due to an increase of oscillator strength. The value of the oscillator strength was one of silicon nanostructures and the reduced mass is taken as a half mass of electron rest mass. For a cluster of 18 atoms, the band gap energy is taken as (1.82) eV with radius 1 nm. The absorption coefficient for nanostructure is given as [26]:

$$\alpha_d \cong \frac{5.4 \times 10^5}{n(\lambda)} \left[\frac{1.24}{\lambda} - 1.82\right]^{\frac{1}{2}} \left(cm^{-1}\right) \tag{20}$$

where λ, is measured in μm and $n(\lambda)$ is the refractive index given by Herzberger's formula.

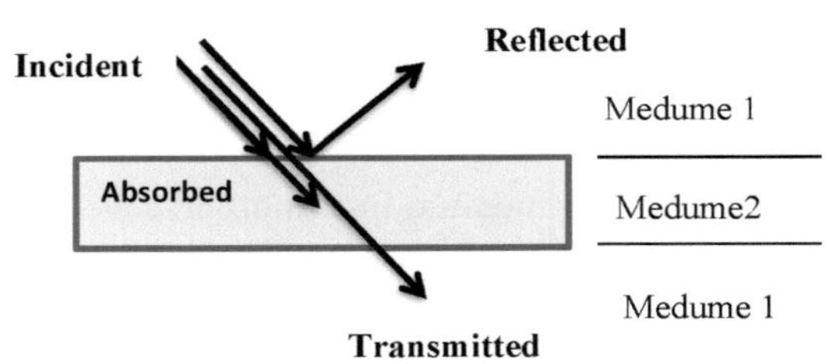

Figure 4.
Diagram of the interaction of light with matter.

$$n(\lambda) = 3.2346 + \frac{0.3698}{\lambda^2 - 0.028} \quad (21)$$

6. Characterization methods

In recent years, it was found that the nanomaterials are very important, and they keep growing in the field of nanoscience and nanotechnology. The researchers used various nanomaterials in the synthesis and application process, due to their potential in the application of science and industry. For example, biocomposite nanomaterials are applied directly and used to replace natural materials to work or to be in contact with the living systems. There are several methods to determine the type of material in the range of nanoscale [27]. Nanoparticle formation is analyzed by using UV-visible spectroscopy and characterization of nanoparticles by SEM, TEM, XRD, FTIR, and EXD. Each method is based on measurements that differ from the other and can be carefully compared. Many of these methods focus on examining particle size at the nanoscale to determine the average particle size of a sample. The role properties of nanoparticles depend on the size and shape, and few particle size distributions of commercial products are narrow in range. In this chapter, the focus is on diagnosing nanocomposites using known techniques that are as follows:

6.1 Transmission electron microscopy (TEM)

A microscopy technique in which a beam of electrons is transmitted through an ultra-thin specimen, and the interaction with the specimen as it passes through it is called transmission electron microscopy (TEM). When the electron beams are transmitted through the specimen as shown in **Figure 5**, the strong interaction between the specimen (atoms) and the electrons duo helps form an image. The image detected by a sensor such as a charge-coupled device (CCD) camera or focused on the device to be an image, such as a fluorescent screen, on a layer of photographic film [28]. Specimens are needed to be very thin, usually below 100 nm in thickness, to achieve good signal-to-noise ratio and sufficient contrast in transmission. Transmission electron microscopy techniques provide two-dimensional images of nanoparticles; these images can be used to produce number-based size distributions, but nanoparticles have all three external dimensions on the nanoscale, and performance properties often depend on their physical-chemical characteristics, that is, size, shape, surface structure, and texture [29]. The perfect sample of transmission electron microscopy for nanoparticle size analysis is one with a large number of individual particles in nanoscale within the desired TEM micrograph field of view, but without excessive agglomeration or bunching of nanoparticles. There are two factors that may have an effect on the TEM grid of the nanoparticle number density: the derivatization efficiency process and the concentration of nanoparticles in solution [30]. Typically, the calculated sizes are expressed as a sphere diameter that the particle has the same projected area as the projected image. Particle size analysis was done using manual or automatic techniques. The first analysis used to get a mean result by obtain a linear dimensional measure of the particle divided by the number of particles, it's usually based on the marking device. To get a clear image, the preparation was elaborated and is slow with few particles being examined [31]. The resolution of image is related to the amplitude and phase alterations in the electron beams that are determined by the contrast transfer function (CTF) as:

$$CTF = A(q)e^{iX(q)} \quad (22)$$

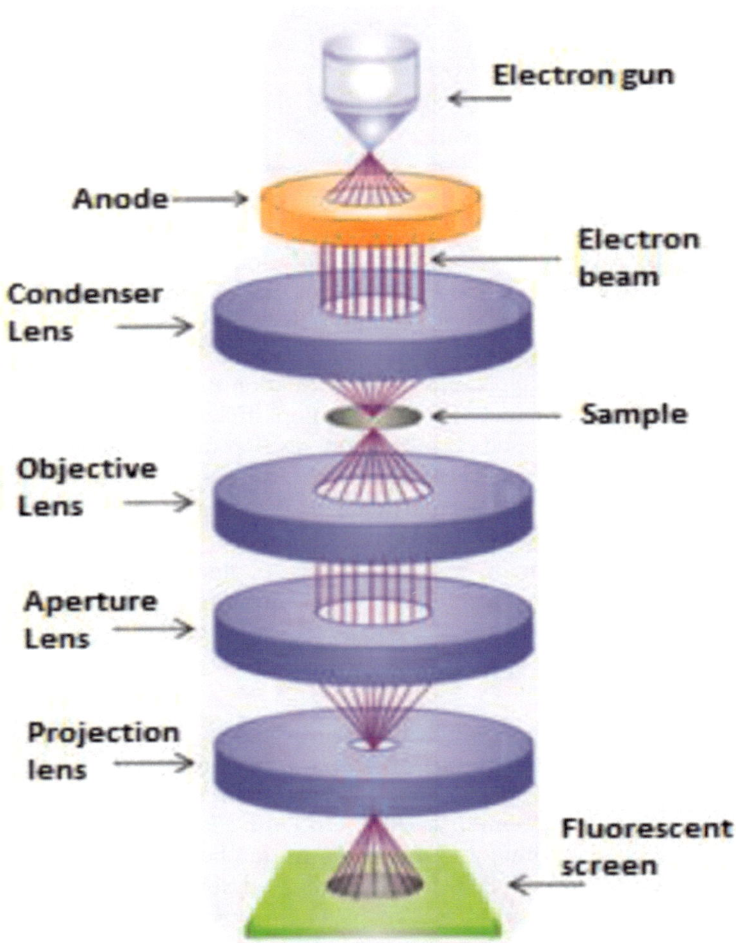

Figure 5.
Schematic form of transmission electron microscopy.

where A (q) describes the diffraction diagram truncation by the aperture of the objective lens and $e^{iX(q)}$ is the phase function, which is described as the distortion of the output wave by the objective lens.

6.2 Scanning electron microscopy (SEM)

The scanning electron microscopy (SEM) is an electron microscope that creates images for the sample surface by scanning it with a high energy stream of electrons [32]. The scheme of SEM is illustrated in **Figure 6**.

The surface morphology of the materials was investigated using scanning electron microscopy (SEM) technique. This technique is different from transmission electron microscopy at site of specimen and intensity of electron beams. For TEM, the electron beam penetrates the sample, but for SEM, the electron beam is incident on the surface of the sample. SEM provides information about surface morphology and composition of materials. There are several advantages for SEM technique in morphological and sizing analysis, but the information is limited for distribution size and true average population. The investigate of solution of nanoparticles with

SEM needed to drying to get on a powder from it before mounted on a sample holder and coating a conductive metal on the surface of sample, such as gold, using a sputter coater. The surface sample is scanned when a high energy stream of electrons is incident on it [33]. The high-resolution magnified images produced when the revealing details about less than 1–5 nm in size and for narrow electron beam yields a characteristic three-dimensional for understanding the surface sample structure.

6.3 X-ray diffraction (XRD)

X-ray diffraction (XRD) is defined as the nondestructive technique that provides detailed information about the crystallographic structure, chemical composition, and physical properties of materials. When the beam of monochromatic incident on the target materials the interaction between them is happened and the scattering of those X-rays from atoms within the target material can be illustrated in **Figure 7**. Bragg's law was used to explain the interference pattern of X-rays scattered by crystals structure the diffraction of

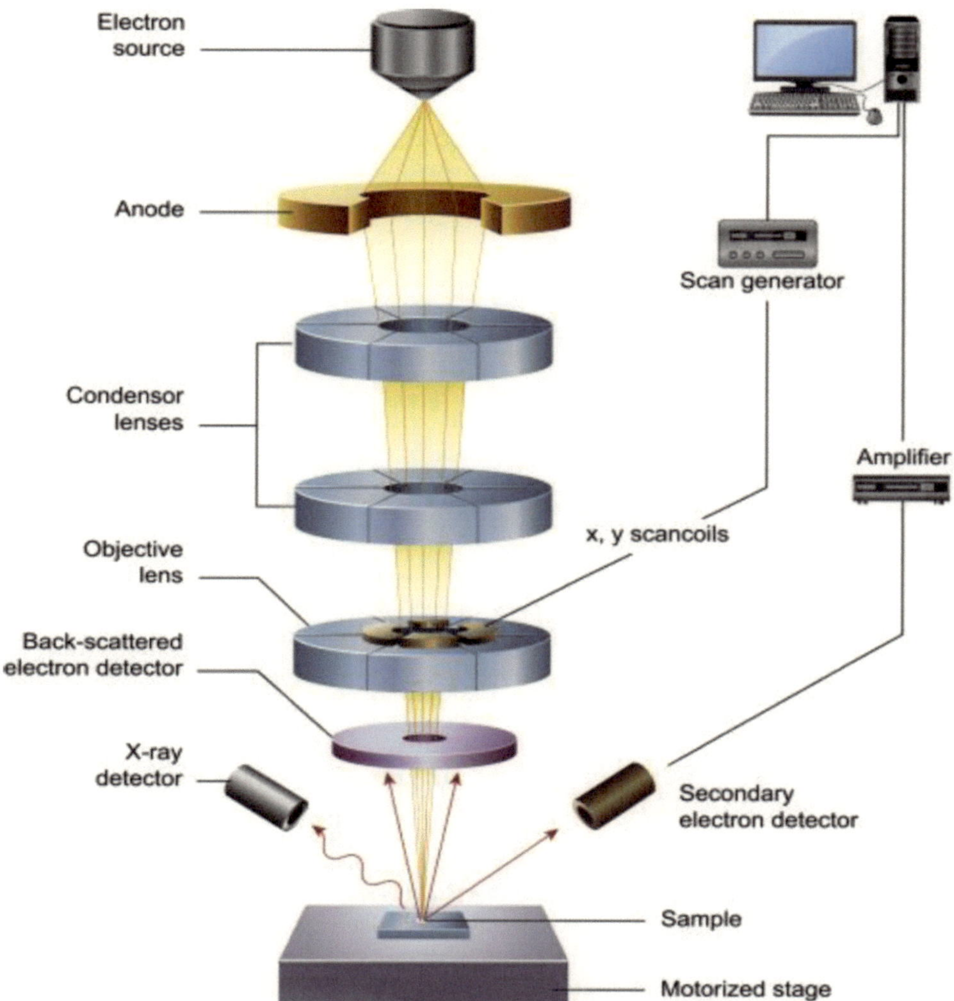

Figure 6.
Schematic form of SEM.

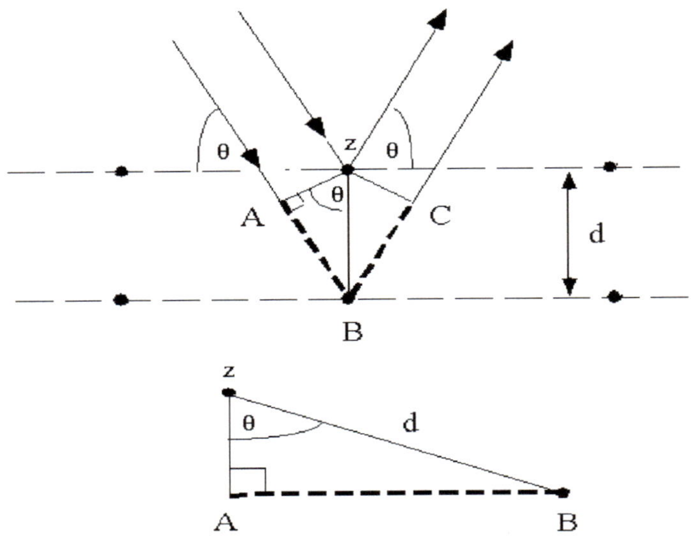

Figure 7.
Schematic diagram of the interaction of the X-ray with mater.

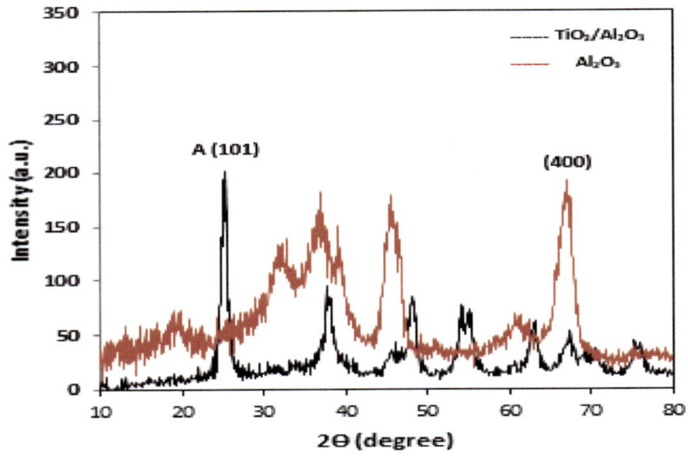

Figure 8.
XRD pattern of Al_2O_3 and TiO_2/Al_2O_3 nanocomposite.

X-rays described by [34]:

$$n\lambda = 2d\,sin\theta \tag{23}$$

where n is an integer, λ is the wavelength of the X-rays, d is the interplanar spacing generating the diffraction, and θ is the diffraction angle.

X-Ray diffraction (XRD) can be considered as a good technique for analyzing the nanostructures, because the width and shape of reflections yield information about the substructure of the materials (sizes of microcrystallites, microdistortions of a lattice, dislocation structures, etc.). There are several approaches to analyze the X-ray diffraction line profiles, with the Scherrer, Williamson-Hall, and Warren-Averbach methods being most widely applied [35].

The Scherrer formula was used by most material scientists as the simplest method of particle size determination. The formula proposed by P. Scherrer in 1918 describes the broadening of diffraction reflection peaks as a function of the average particle size D [16]:

$$\beta = k\frac{\lambda}{D\ cos\varphi} \tag{24}$$

where k is equal to:

$$k = 2\sqrt{(ln\,2)/\pi} \approx 0.94 \tag{25}$$

and k is the shape factor, λ is the incident x-ray wavelength (0.15040 nm for CuK), β is full width at half maximum (FWHM), and φ is diffraction angle at maximum intensity peak.

XRD pattern of Al_2O_3 and TiO_2/Al_2O_3 nanocomposite is shown in **Figure 8**. The diffraction Al_2O_3 peaks can be well indexed to pure α-Al_2O_3 (JCPDS Card no. 880826). The appearance of diffraction peaks in TiO_2/Al_2O_3 XRD pattern corresponding to (101) and other planes is in good agreement with the standard XRD peaks of Anatase TiO_2 (JCPDS Card No. 040477). The average crystal sizes of nanocomposite materials for TiO_2/α-Al_2O_3 (21.4 nm) are larger than Al_2O_3 (8.1 nm), which leads to get a good mix of NPS.

7. Nanotechnology applications in the environment

Nanoparticles that are produced deliberately using specific processes are called engineered or manufactured nanoparticles, for example, fullerenes and CNTs. With regard to environmental issues, the system of one dimensional (1D), thin films, or surfaces of two dimensional (2D), this can be used in applications of electronics, chemistry, and engineering as thin films at the range of sizes (1–100 nm) or monolayer in the field of solar cells or catalysis. These thin films are inserted in different technological applications, including development of a new generation of environmental sensing systems, chemical and biological sensors, fiber-optic systems, and magneto-optic and optical device.

The sun sends an infinite light free from environmental pollution and noise is a renewable source of energy. The energy drawn from the sun can easily compensate for nonrenewable sources of energy such as fossil fuels and petroleum deposits on the earth. The solar cells have passed through a large number of improvement steps from one generation to another, because of their importance for the generation of alternative energy [36].

7.1 Solar cell

Photovoltaic (PV) is related to the devices such as solar cell that directly converts sunlight into electricity. The solar cell is the elementary building block of the photovoltaic technology. Silicon is one of the most common semiconductor materials that is used to make solar cells. One of the most common properties of semiconductors that makes them most useful is that their conductivity may easily be modified by introducing impurities into their crystal lattice. There are several types of solar cells, and they are either cut from a single crystal rod or from a block composed of many crystals and are correspondingly called monocrystalline or multicrystalline silicon solar cells and nanocrystal-based solar cells [37]. Most solar cells are fundamentally large areas of p-n junctions. When light shines on them, they can generate current and voltage, the photons produce electron-hole (e–h) pairs, and the dipole electric field provides for a separation of these charges. The reason this can happen is because of the "built-in" electric field at the junction of the p-type and n-type material [36]. The junction between them creates a charge separation region with a strong dipole electric field.

The current-voltage (I-V) characteristics of photovoltaic cell are illustrated in **Figure 9**, which operates under normal conditions. The power curve is obtained when a solar cell produced power and then the current and voltage ($I \times V$) are the products. Most solar cells behave as a diode in the dark, admitting a much larger current under forward bias (V > 0) than under reverse bias (V < 0). For an ideal diode, the dark current density varies as:

$$J_{dark} = J_o\left(e^{qV/K_BT} - 1\right) \quad (26)$$

where J_o is a constant. Thus, the net current flowing in a circuit powered by a solar cell is:

$$J(V) = J_{sc} - J_{dark} \quad (27)$$

$$J(V) = J_{sc} - J_o\left(e^{qV/K_BT} - 1\right) \quad (28)$$

where I_{sc} is the current of short-circuited and V_{oc} is the voltage of open circuit.

The maximum power is obtained when ($V=V_m$ and $I=I_m$) and the fill factor (FF) is defined by the ratio [38]:

$$FF = \frac{V_m I_m}{V_{sc} I_{sc}} \quad (29)$$

where V_m, I_m is the maximum voltage and current, respectively.

Quantum efficiency (QE) is the ratio of the number of charge carriers collected by the solar cell to the number of photons of a given energy incident on the PV device [39]:

$$QE = \frac{P_{out}}{P_{in}} = \frac{P_m}{P_s} \quad (30)$$

Where is the maximum power and $P_m = V_m I_m$,Ps is incident light power.

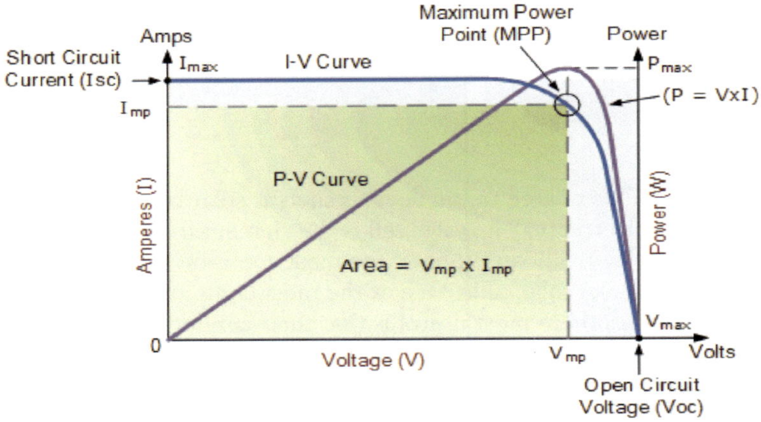

Figure 9.
(I-V) Characteristics of a typical PV cell.

7.2 Nanocoatings

Coating is defined as a coherent layer formed from a single or multiple applications of coating materials to a substrate. According to the existing standard, coating material is a material in liquid, paste, or powder form that, when applied, forms a protective and decorative coating. Some nanomaterials are suitable for use in transparent coating systems. In addition, the transparency of these nanomaterials such as TiO_2 nanoparticle in visible light makes it possible to create novel additives introducing new properties to otherwise nontransparent coatings. The choice of the manufacturing process depends on the specific application and the specific application requirements of the coating. The sol-gel process may offer several advantages to manufacturers: the manufacturing process is shorter, runs at lower temperatures, and consumes less energy.

The properties of Titanium dioxide (such as high photocatalytic efficiency, chemical stability, low toxicity, and low cost) made it most thoroughly used from other materials. Also, self-cleaning paints with other metal oxides like ZnO have been reported [40]. The self-cleaning researchers are mostly about air pollution and environmental contamination in buildings especially on indoor and outdoor building surfaces. The wide range of applications of self-cleaning was necessary to focus on various materials for different purposes [41]. Although the properties of surfaces of self-cleaning are complex, however, it is related to several of their surface characteristics. At the beginning, the surfaces are superhydrophilic and water droplets are spread across the surface making it easier to wash off solid material. Then any organic material coating on solid particles will react with these surfaces by photocatalytic reactions to allow them to fall or wash off more readily. The surfaces of TiO_2 have very high electroconductivity. A surface with high electroconductivity provides antistatic properties repelling charged particles and preventing their accumulation on the surface. In addition, the waterborne paint is prepared by the mill base for the pigment dispersion in water, auxiliary solvents, etc. Then the mill base is blended with the binder (polymer latex) and the paint is obtained [40].

7.3 Sonocatalyst

The chemical effects of ultrasound are not derived from a direct coupling of the acoustic field with chemical species on a molecular level. Instead, sonochemistry and sonoluminescence derive principally from acoustic cavitation [42]. In the past decade, the expansion of the sonoelectrochemistry has become increasingly important. The variety of induced effects on electrochemistry processes by ultrasound waves can be attributed to the generation, growth, and collapse of microbubbles in the electrolyte. There is a growing interest of the application of the sonoelectrochemistry in environmental remediation and in the preparation of nanopowders [43]. Ultrasounds have a wide range of uses in the development of applications of nanoparticle solutions for different chemical compounds. The effect of ultrasonic energy breaks the chemical bonds of compounds [44]. The ultrasonic catalytic degradation method has been widely used in wastewater treatment because of its many excellent properties; these are simple equipment, have high efficiency and stable operation, are safe, and cause no secondary pollution. Improvement of ultrasonic catalysis process is known as a sonocatalyst [45]. It has received great attention as a useful and promising method for mineralizing organic pollutants, for example, synthetic dyes in aqueous media. In this process, water molecules are used to produce hydroxyl radicals, which are very reactive and non-selective oxidants and are capable of decolorizing and mineralizing dyes to CO_2 and H_2O. The oxidation processes of a metal oxide semiconductor are advanced through ultrasonic treatment on surface [46]. The presence of semiconductor particles (i.e., TiO_2,

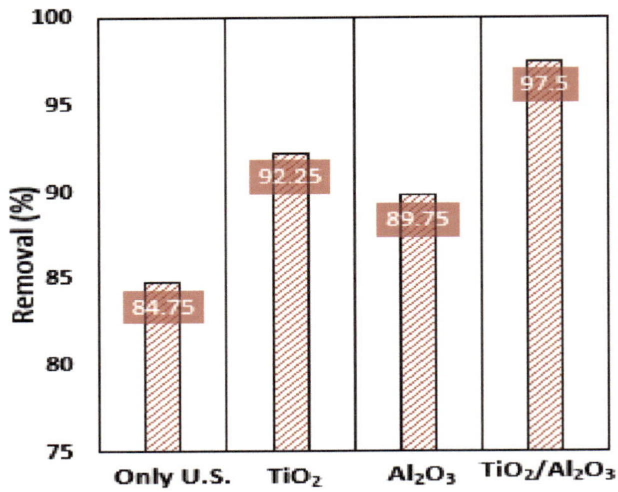

Figure 10.
The sonocatalysis effect on MB decolorization.

ZnO) enhanced the process of breaking up the microbubbles created by the ultrasound irradiation into smaller bubbles, and these proses will be increasing the quantity of high of high temperatures and pressures, this leads to produce additionally amount of hydroxyl radicals which will attack the pollutant and resulting in degradation of the pollutant. Sonochemical decolorization of dyes under initial concentrations using ultrasonic processor represented by the effect of nanocomposite of Al_2O_3, TiO_2, and TiO_2/Al_2O_3 on the decolorization of methylene blue dye was clear as shown in **Figure 10**. The increase in the decolorization of dye in the presence of nanoparticles due to these nanoparticles act as catalysts that increase the number of nucleation of the cavity and improve the rate of dissociation of water into highly reactive hydroxyl radicals (·OH). TiO_2/Al_2O_3 nanocomposites show highly removal of Methylene blue dye than other Sonocatalysts due to highly dissociation rates H_2O molecules that yields more free radical generated, thereby increasing the rate of degradation of the organic compounds [16].

8. Conclusion

Nanomaterials can be used in different applications such as in medicine, electronic device, sunscreens, military applications, photovoltaic cells, paints, catalysts, etc. Some of these do not have an effect on the environment, while others have an effect on it. In this chapter, the focus of our attention was on the applications that do not affect the environment and improve it, so the important property that surface to volume ratio of nanomaterials increases with decreases particle size. To do that, the processes of preparing nanoparticles are physical and chemical methods, and the sol-gel process is basic to prepare nanomaterials in chemical methods such as TiO_2, Al_2O_3, and $TiO_2/\alpha\text{-}Al_2O_3$ because it can be used at low temperature and short time. These can be used in solar cells to produce clean energy, nanotechnologies in coatings, and sonochemical decolorization of dyes.

Author details

Ali Salman Ali
Physics Department, College of Science, Al-Muthanna University, Al-Muthanna, Iraq

*Address all correspondence to: alisbasrah@yahoo.com

References

[1] Zhang ZY, Lagally MG. Atomic processes in the early stages of thin-film growth. Science. 1997;**276**:377-383

[2] Ngo C, Van de Voorde M. Nanotechnology in a Nutshell. 2014. DOI: 10.2991/978-94-6239-012-6

[3] Srinivasan S, Kannan AM, Kothurkar N, Khalil Y, Kuravi S. Nanomaterials for energy and environmental application. Journal of Nanomaterials. 2016;**2015**:2. Article ID: 979026

[4] Tala-Ighil R. Handbook of Nanoelectrochemistry. 2015. DOI: 10.1007/978-3-319-15207-3_26-1

[5] Claire A, Grégory B, Garidel-Thoron CDE. Painting the future: AMIPAINT nanomaterials and the safer-by-design approach for new markets. SSRN Electronic Journal. 2017. DOI: 10.2139/ssrn.3033840

[6] Hitesh S. Material today. Science Direct. 2018;**5, 1**(2):6227-6233

[7] Filipponi L, Sutherland D. Fundamental Nano-effects. Denmark: Inano, Aarhus University; 2010

[8] Wang ZL, editor. Characterization of Nanophase Materials. New York: Wiley-VCH; 2000

[9] Roduner E. Size matters: Why nanomaterials are different. Chemical Society Reviews. 2006;**35**:583-592

[10] Arole VM, Munde SV. Fabrication of nanomaterials by top-down and bottom-up approaches: An overview. JAAST: Material Science. 2014;**1**(2): 89-93

[11] Guo Z, Tan L. Fundamentals and Applications of Nanomaterials. ARTECH HOUSE 685 Canton Street, Norwood, MA 02062 P.95; 2009

[12] Caterina M. Bottom-up approaches for organizing nanoparticles with polymers, Thèse No. 3092, laurea in fisica. Italie: Università degli Studi di Firenze; 2004

[13] Helmer F, Anja OS. Synthesis; sol-gel. Uni. Of Oslo, Dept. of Chemistry, Lectures at CUTN; Spring 2016

[14] Bessekhouad Y, Robert D, Weber JV. Preparation of TiO_2 nanoparticles by Sol-Gel route. International Journal of Photoenergy. 2003;**05**:153-158

[15] Farahmandjou M, Golabiyan N. Synthesis and characterization of alumina (Al_2O_3) nanoparticles prepared by simple sol-gel method. International Journal of Bio-Inorganic Hybrid Nanomaterials. 2016;**5**(1):73-77

[16] Ali AS, Mohammed AJ, Saud HR. Hydrothermal synthesis of TiO_2/Al_2O_3 nanocomposite and its application as improved sonocatalyst. International Journal of Engineering & Technology. 2018;**7**(4.37):22-25

[17] Christian P, Von der Kammer F, Baalousha M, Hofmann T. Nanoparticles: Structure, properties, preparation and behaviour in environmental media. Ecotoxicology. August 2008. DOI: 10.1007/s10646-008-0213-1

[18] Guo D, Xie G, Luo J. Mechanical properties of nanoparticles: Basic and applications. Journal of Physics D: Applied Physics. 2014;**47**:013001 (25pp)

[19] Hadjipanayis GC, Siegel RW, editors. Mechanical properties of nanophase materials. Nanophase Materials. Kluwer Academic Publishers; 1994. pp. 233-261

[20] Available from: https://www.sciencedirect.com/referencework/9780080431529

[21] Liang T. A short review on mechanical behavior of nanocrystalline materials. International Journal of Metallurgy and Metal Physics. 2017;**2**: 008

[22] Ederth J. Uppsala: Department of Materials Science, Uppsala University; 2003. ISSN 1104-232X, ISBN: 91-554-5499-6

[23] Ali S. Ali, Kawakib J. Majeed. 1st Annual International Interdisciplinary Conference, AIIC 2013, 24–26 April, Azores, Portugal

[24] McLachlan DS, Sauti G, Corporation HP. The AC and DC conductivity of nanocomposites. Journal of Nanomaterials. 2007. p. 9. Article ID 30389

[25] Ali AS, Mohammed AJ. The optical properties of thin films of polymer PMMA/MR-Eosin prepared by casting methods. Journal of International Academic Research for Multidisciplinary. February 2017;**5**(1)

[26] Kulkarni A, Guney D, Vora A. Optical absorption in nano-structures: classical and quantum models. ISRN Nanomaterials. 2013. p. 7. Article ID: 504341

[27] Ghoranneviss M, Soni A, Talebitaher A, Aslan N. Nanomaterial synthesis, characterization, and application. Journal of Nanomaterials. 2015. p. 2. Article ID: 892542

[28] Heera P, Shanmugam S. Nanoparticle characterization and application: An overview. International Journal of Current Microbiology and Applied Sciences. 2015;**4**(8):379-386

[29] Rice SB et al. Particle size distributions by transmission electron microscopy: An interlaboratory comparison case study. Metrologia. 2013;**50**(6):663-678

[30] Bonevich JE. Nanotechnology Characterization Laboratory, National Cancer Institute- Frederick, MD 21702, (301), 846-6939; 2010

[31] Akbari B, PirhadiTavandashti M, Zandrahimi M. Particle size characterization of nanoparticles: A practical approach. Iranian Journal of Materials Science and Engineering. 2011;**8**(2)

[32] Joshi M, Bhattacharyya A, Wazed Ali S. Characterization techniques for nanotechnology applications in textiles. Indian Journal of Fibre and Textile Research. 2008;**33**:304-317

[33] Jores K et al. Investigation on the stricter of solid lipid nanopartuicles andoil-loaded solid nanoparticles by photon correlation spectroscopy, fieldflowfractionasition and transmission electron microscopy. Journal of Controlled Release. 2004;**17**:217-227

[34] Bunaciu AA, Udristioiu EG, Aboul-Enein HY. Critical reviews in analytical chemistry. 2015;**45**:289-299

[35] Dorofeev GA et al. Determination of nanoparticle sizes by the Xray diffraction method. Colloid Journal. 2012;**74**(6):678-688

[36] Sharma S, Jain KK, Sharma A. Solar cells: In research and applications: A review. Materials Sciences and Applications. 2015;**6**(12):1145-1155

[37] Piebalgs A, Potočnik J. Photovoltaic Solar Energy. Development and Current Research, European Communities; 2009. DOI: 10.2768/38305

[38] Yuan C. Development of nanoparticle sensitized solar cells. In: Doctoral Thesis in Theoretical Chemistry & Biology, School of Biotechnology. Stockholm, Sweden: Royal Institute of Technology; 2013

[39] Ananda W. International Symposium on Electrical and Computer

Engineering Conference Paper. IEEE; 2017. DOI: 10.1109/QIR.2017.8168528

[40] Bonnefond A, González E, et al. Stable photocatalytic paints prepared from hybrid core-shell fluorinated/acrylic/TiO_2 waterborne dispersions. Crystals. 2016;**6**:136. DOI: 10.3390/cryst6100136

[41] Chermahini SH et al. New trends in self-cleaning materials for different purposes. International Journal of Advances in Civil Engineering. 2018;**1**(1)

[42] Kis-Csitári, Kónya Z, and Kiricsi I. Sonochemical synthesis of inorganic nanoparticles. Springer Science + Business Media B.V.; 2008. pp. 369-372

[43] Sáez V, Mason TJ. Sonoelectrochemical synthesis of nanoparticles. Molecules. 2009;**14**: 4284-4299. DOI: 10.3390/molecules14104284

[44] Vasile-Sorin M, Aloman A. Obtaining silver nanoparticles by sonochemical methods. UPB Scientific Bulletin, Series B: Chemistry and Materials Science. 2010;**72**(2)

[45] Song S, Hao C, Zhang X, Zhang Q, Sun R. Sonocatalytic degradation of methyl orange in aqueous solution using Fe-doped TiO_2 nanoparticles under mechanical agitation. Open Chemistry. 2018;**16**:1283-1296

[46] Khataeea A, Kayanc B, Gholamia P, Kalderisd D, Akay S. Sonocatalytic degradation of an anthraquinone dye using TiO_2-biochar nanocomposite. Ultrasonics Sonochemistry. 2017;**39**: 120-128

Chapter 3

Biological Synthesis of Nanoparticles Using Endophytic Microorganisms: Current Development

Omar Messaoudi and Mourad Bendahou

Abstract

Nanotechnology is a new emerging interdisciplinary approach created by pairing of engineering, chemical, and biological approaches. This technology produces nanoparticles using different methods of traditional physical and chemical processes; however, the outlook in this field of research is to use ecofriendly, nontoxic, and clean methods for the synthesis of nanoparticles. Biological entities, such as plants, bacteria, fungi, algae, yeast, and actinomycetes, are the best candidate to achieve this goal. Among the biological route, those involve endophtic microorganisms to reduce metallic ions into nanoparticles. This method is considered as an attractive option and can open a new horizon on the interface of biology and nanotechnology. The present chapter highlights the latest research about endophytic microorganisms and their application in the synthesis of nanoparticles, as well as the mechanisms involved in the formation of nanoparticles.

Keywords: endophyte microorganisms, green nanotechnology, nanoparticles

1. Introduction

Nanotechnology is a new emerging interdisciplinary approach of created by pairing of biotechnology, and nanotechnology [1]. This new technology produced nanoparticles of various types (silver, copper, zinc, gold, etc.) at the nanoscale level (less than 100 nm). Three different methods can be employed for the synthesis of nanoparticles, including, chemical, physical and biological methods. These three methods follow either the bottom-up approach, or the top-down approach for the synthesis of nanoparticles [2].

The outlook in this field of research is to use ecofriendly, nontoxic and clean method for the synthesis of nanoparticles [3]. The chemical and physical methods are generally expensive and associated with destructive effects on the environment and human health [4]. In order to counter those limitations, one of the proposed solution is the application of a novel route for producing nanoparticles based on bottom-up method, called 'green synthesis', which is regarded as an important tool and gaining great attention in current research. This method is based on the utilization of natural resource, such as plants, fungi, bacteria, actinomycetes, yeast and algae, to reduce the metal ions into metallic nanoparticles [5].

The green synthesis of nanoparticles offers a set number of benefits compared with physical and chemical methods, since this method is cost-effectively, eco-friendly, uses less energy and can provide nanoparticles with better defined size and morphology, with a great compatibility for pharmaceuticals, medical, agronomical and environmental applications [6].

Microbial-mediated biosynthesis of nanomaterials is one of the promising biological-based nanomanufacturing process [7]. Microorganisms can produce nanoparticles by intracellular or extracellular synthesis, according to the location where nanoparticles are formed, through enzymes or biomolecules generated by the cell activities [8]. The use of microorganisms offers different advantages over the biosynthesis of nanoparticles by plants and algae, since microorganism can be easily scale-up, and they offert the possibility to changing culture condition to obtained nanoparticles with desired shape and sizes [9].

One approach that shows immense potential is based on the biosynthesis of nanoparticles using endophytic microorganisms, which is considered as a new potential source, under explored [10]. In this chapter, we present, the latest research about nanoparticles from endophytic microorganisms.

2. Endophytic microorganism: bacteria and fungi

"Endophytes" is a Greek word that mean "within plant", this term is used for microorganisms (bacteria or fungi) that dwell within plant tissues, without causing any disease, infection, or damage to the plant tissues [11]. Every plant host, intercellularly and/or extracellularly, in various spaces of plant parts including roots, leaves, stems, flowers, and seeds, one to more endophytes microorganisms [12]. To date, endophytes microorganism has been found in all plant species that exist on the earth (nearly 390,000 plants) [13]. Mutualist is the most common relationship between plants and endophytes, however, in some cases and under some conditions, the endophytes can behave as opportunistic pathogens [14].

To have a stable symbiotic relationship, the plant host provides to endophytes the necessary organic nutrient, generated through photosynthesis, for growth and multiplication [14]. On the other side the endophytes offer different beneficial effects to the host plant, this including: (i) nutrient assimilation: by synthesis of iron (Fe)-sequestering siderophores, fixation of atmospheric nitrogen, solubilization of minerals such as phosphorus [15]. (ii) Stimulation of plant growth: by secretion of plant growth regulators (PGRs), such as auxin, cytokinin, ethylene and gibberellin [16]. (iii) Protection of host plants from attack of pathogens microorganisms and insects: through secretion of various bioactive secondary metabolites as well as lytic enzymes [17].

The endophytes microorganisms can be acquired directly from the environment (horizontal transmission), or are vertically transmitted from generation to generation via seed [18]. The majority of endophytes are acquired via the first mechanism of transmission, this was confirmed through the study of the diversity of microorganisms in seeds and seedlings, raised under sterile conditions, which are typically lower than the diversity of microorganisms in plants grown in soil [19].

Endophytes are studied under two categories, bacterial endophytes and fungal endophytes [20]. The structure of the microorganism communities resides inside the plants, depends on several factors, including, the nature of soil and the plant host species [21]. To study the composition in microorganisms of endophytes, the culture-dependent methods do not allow a complete overview of the endophytic population, because the uncultured microorganisms cannot be recovered and identified using this method. However, the use of molecular approaches, including high throughput techniques of next generation sequencing (NGS), confers a rapid analysis of the composition and diversity of plant microbial endophytes communities [22]. According to the study of Hardoim et al., 2014 [14], which analyze the sequences of 16 s DNAr assigned to endophytic bacteria strains, including cultured and uncultured bacteria, he found that, 96% of analyzed sequences belong to four different cultured phyla, which is reported to be dominant in the plant environment, including: 54% *Proteobacteria, 20% Actinobacteria, 16% Firmicutes,* and 6% *Bacteroidetes*. However, 19 phyla belong to the non cultured bacteria. Furthermore, 50% of the analyzed sequences, which are the predominant endophytes strains, belong to the genera, *Pseudomonas, Enterobacter, Pantoea, Stenotrophomonas, Acinetobacter,* and *Serratia*, all these genera are member within the class of *Gammaproteobacteria (Proteobacteria phylum).* Other genera are also well represented whithin endophytic bacteria population, this including *Streptomyces, Microbacterium, Mycobacterium, Arthrobacter*, as well as *Bacillus, Paenibacillus, and Staphylococcus*.

Endophytic fungi are ubiquitous in plants and are mainly members of *Ascomycota* or their mitosporic fungi, as well as some taxa of *Basidiomycota, Zygomycota,* and *Mucoromycota* [23]. Li et al. [24], examined endophytic fungi associated with the stem and root of 10 halophytic species colonizing the Gurbantonggut desert, they obtained 36 endophytic fungal taxa, dominated by *Alternaria eichhorniae, Monosporascus ibericus,* and *Pezizomycotina* sp. 1. However, a total of 56 endophytic fungi was isolated from leave and root segments of *Salvia abrotanoides* at the three sites by Teimoori-Boghsani et al. [25]. The isolated strains belong to 16 different fungal genera, this including: *Penicillium, Paraphoma, Phaeoacremonium, Talaromyces, Aspergillus, Psathyrella, Trichoderma, Alternaria, Thielavia, Acremonium, Fusarium, Talaromyces, Coniolariella, Paecilomyces, Simplicillium, and Monocillium.* Among the obtained strains, only two isolates were recovered from the plant's leaves (*Thielavia microspore* and *Aspergillus* sp.), while the remaining isolates were obtained from root samples.

3. The green nanotechnology

Nanotechnology is a rapidly growing field of science, and can be defined as the manipulation of materials at the nanometer scale or one billionth of a meter. It's become an integral part of the biotechnology and regarded as one of the key technologies [26].

Nanotechnology produces materials which have one dimension less than 100 nm at least, these materials, called nanoparticles, can be produced using different metals, such as: gold (Au), silver (Ag), copper oxide (CuO), zinc oxide (ZnO), iron (Fe2O3), palladium (Pd), platinum (Pt), nickel oxide (NiO), magnesium oxide (MgO), selenium (Se) and titanium dioxide (TiO2) [27].

The synthesis of nanoparticles is based on two approaches: (1) top-down approach and (2) bottom-up approach (**Figure 1**) [28]. The first approach (top-down approach) is destructive method, based on the decomposition of larger molecule into smaller units, these unit are then converted into appropriate nanoparticles. Several physical methods are applied in this case: mechanical milling, chemical etching, sputtering, laser ablation electro-explosion [29]. The second approach (bottom-up approach), is employed in reverse to the first approach, in fact, in this case, nanoparticles are formed when atoms are self assemble together [30]. The synthesis of nanoparticles using this approach, can be carried

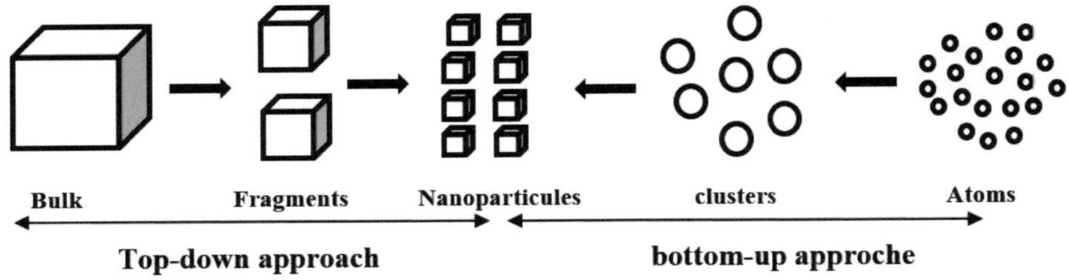

Figure 1.
Top-down and bottom-up approach.

out by several physical and chemical methods including: spinning, template support synthesis, plasma or flame spraying synthesis, laser pyrolysis, CVD, atomic or molecular condensation [31]. Biological routes can also be applied to reduce metallic ions into neutral atoms (zero valent atoms) for synthesis of nanoparticles with bottom-up approach, this method is so called green nanotechnology, in this case several biological sources, available in nature, are involved, such as: (i) utilization of microorganism (bacteria, fungi); (ii) utilization of plant extracts; (iii) utilization of microseaweeds; (iv) using enzymes and biomolecules [32, 33].

Biological agents involved in green nanotechnology offer many benefits as compared with physical and chemical syntheses, in fact, these techniques are costly, requires higher utilization of energy, and utilize toxic chemicals that may have a disastrous effect on the environment [34]. In contrast, biological approach has several edges over chemical and physical methods for synthesis of nanoparticles, as it is low cost, eco-friendly, non-toxic, clean and can be scaled up to larger-scale synthesis with ease [35].

Biological nanoparticles, synthesized using different metal, have been applied in many fields, in fact, the silver nanoparticles are widely used in medical fields, for example Al-Sheddi et al. [36], show the potential of silver nanoparticles synthesized using an extract of *Nepeta deflersiana* against Human Cervical Cancer Cells (HeLA). However, Soliman et al. [37] indicate that the silver nanoparticles synthesized by the pink yeast, *Rhodotorula* sp. ATL72, isolated from salt marches near mediterranean sea, Egypt, exhibited strong antimicrobial activity against a wide range of Gram positive and Gram negative bacteria as well as fungi with low MIC value. Moreover, zinc and titanium nanoparticles are generally used in cosmetics fields [38]. Biological nanoparticles can also apply as sensors for various biomolecules related to environmental factors and agriculture, as well as they can also use for gene delivery and cell labeling in plants and in medicine [39].

4. Mechanisms of nanoparticle biosynthesis by microorganisms

Although, the number of studies which elucidate the green synthesis of nanoparticles using microorganisms, there is a little work about the mechanism and the biochemical pathway involved behind the synthesis of metal nanoparticles.

Intra and extra cellulary microbial enzymes and secondary metabolites secreted by microorganisms, play a key role in the reduction of metal ions into their respective nanoparticles. In fact, It has been found that the microorganisms when are exposed to metal ion solution, they are responding to this environmental stress by the secretion of enzymes and biomolecules that possess a reducing potential of metal salts, consequently the metal ions are detoxified to less toxic metal nanoparticles [5].

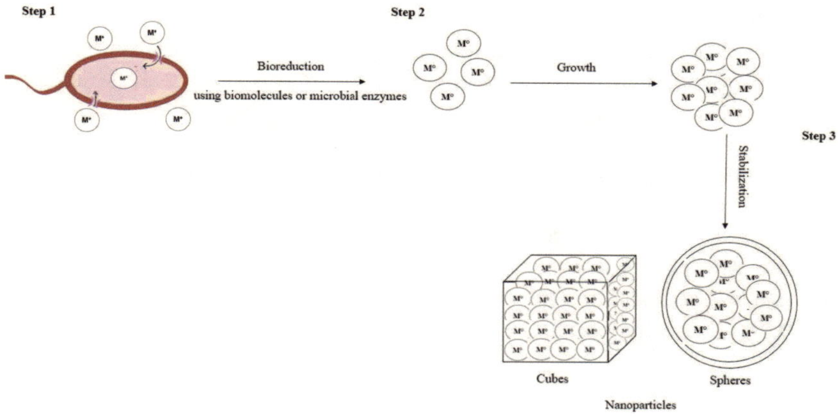

Figure 2.
Mechanisms of nanoparticle synthesis.

Three steps are involved in the biosynthesis of nanoparticles by microorganisms (**Figure 2**):

- In the first step, metallic ions are captured on the surface of microbial cells via electrostatic interaction with the negatively charged cell wall, or they are absorbed inside the microbial cells, through cationic membrane transport systems that normally transport metabolically important cations [5, 40].

- In the second step, metallic ions (M+) are bioreduced into zero-valent metals (M°). This reaction can be catalyzed by: (i) the active groups, such as the hydroxyl group (C-OH) or the ionized carboxyl (COO-) group, of biomolecules biosynthetized by the microorganisms having reduction capabilities., or (ii) or by microbial enzymes, such as, NADH-dependent nitrate reductase, which catalyze the reduction of silver ions to silver nanoparticles at pH 7.2, using NADH as electron source and 8-hydroxyquinoline as electron shuttle [41, 42]. As results of this reduction, the metal ions are changed from their mono- or divalent oxidation states to reduced metal ions (zero-valent states). Afterward, the nanoparticles joint to form different morphology shapes such as, spheres, hexagons, triangles, cubes, ovale, etc. [43].

- The third step corresponding to the stabilization of nanoparticles with capping agents, to prevent further growth and agglomeration and controlling the shape and size of the biosynthesized nanoparticles [5].

The size of nanoparticles biosynthesize by endophytic microorganisms affect the activity, it has been proved that nanoparticles with small size provide great surface/volume ration and guarantee a good activity [44]. Different physico-chemical parameters should be controlled and optimized, such as, temperature, pH, metal salt concentration, incubation period, agitation, nature and concentration of carbon and nitrogen source in culture media, to producing homogeneous nanoparticles in size and shape, with satisfied activity [38].

5. Nanoparticles synthetized by endophytic microorganisms

Biological methods are being a popular trend in the synthesis of metal nanoparticles. Among them, those involving saprophytic microorganisms (bacteria and fungi), which are able to turn the metal ions, from their environment, into metallic nanoparticles through enzymes and secondary metabolites generated by the cell activities. This process provides greater stability and appropriate dimensions of synthesized nonoparticules [37].

Compared with saprophytic microorganisms, the application of endophytic microorganisms has emerged as a novel research area for the green synthesis of nanoparticles. This field of research can open a new horizon, on the interface of biology and nanotechnology, for novel nanomaterials with diverse applications [45].

Different endophytic microorganisms, including fungi, bacteria and actinomycetes, can be used for the biosynthesis of nanoparticles from different metal, such as silver, gold, zinc, copper, etc. **Table 1** summarizes the recent researches in this field.

5.1 Nanoparticles synthesized by endophytic bacteria

Some endophytic bacteria, have developed a specific defense mechanism to overcome toxicity of metal ions, this mechanism is based on the precipitation of ions metals at the nanometer scale to produce nanoparticles [63]. It was observed that some of endophytic bacteria could survive and grow even at high metal ion concentrations. Bacteria possess such remarkable ability to reduce metal ions into nanoparticles, can be a good candidate for nanoparticles synthesis [64].

Ibrahim et al. [46, 47] reported the isolation of *Bacillus siamensis* C1 from *Coriandrum sativum* and *Pseudomonas poae* CO from *Allium sativum*, both strains produce silver nanoparticles with spherical shape and exhibited potential antibacterial, antibiofilm and antifungal activity.

Gold nanoparticles with spherical form and size range from 5 to 50 nm, has been successfully synthesized by the endophytic bacteria, *Pseudomonas fluorescens* 417, isolated from the plant, *Coffea arabica*. The synthesized gold nanoparticles show bactericidal activity against a panel of clinically significant pathogens [49]. The same author, Syed et al. [48], use the strain *Aneurinibacillus migulanus,* isolated from surface sterilized inner leaf segment of *Mimosa pudica,* for the biosynthesis of silver nanoparticles with different shapes, including, spherical, oval, cubic and triangular shapes. The particle size has been determined by Dynamic Light Scattering (DLS) method, and revealed average size of 24.27 nm. The bactericidal activity of the biosynthesis silver nanoparticles indicates interesting activity against both Gram-positive and Gram-negative pathogenic bacteria. The highest activity was observed against *Pseudomonas aeruginosa*, which is considered as clinically important bacteria.

5.2 Nanoparticles synthesized by endophytic fungi

In recent years, the utilization of endophytic fungi for the production of metallic nanoparticles has attracted more attention, due to their metal toleration, metal uptake and accumulation capability [65]. Compared with the other microorganisms, fungi are good machines for the synthesis of any type of metallic nanoparticles, and can provide a several advantages, such as: (i) Easy for isolation from soil or plants, compared with rare bacteria and actinomycetes, which required specific enrichment methods for isolation [56]. (ii) Secrete large amounts of metabolites and extracellular enzymes, which facilitate the reduction of metal ions into nanoparticles. (iii) Easy to scale-up, since they have a rapid growth [66] (iv). Most of the fungi have a large range of growth for pH, temperature and Nacl, which facilitate the change of culture conditions in order to produce homogeneous nanoparticles [67].

Plants	Endophytes	Shapes	Size	Types of NPs	Activity	References
Coriandrum sativum	*Bacillus siamensis* C1	Spherical	25–50 nm	Silver	Antibacterial	[46]
Allium sativum	*Pseudomonas poae* CO	Spherical	19.8–44.9 nm	Silver	Antifungal	[47]
Mimosa pudica	*Aneurinibacillus migulanus*	Spherical, oval, cubic, triangular	~ 24.27 nm	Silver	Antibacterial	[48]
Coffea arabica	*Pseudomonas fluorescens 417*	Spherical	5–50 nm	Gold	Antibacterial	[49]
Raphanus sativus	*Alternaria* sp.	Spherical	4–30 nm.	Silver	Antibacterial	[35]
Taxus baccata	*Nemania* sp.	Spherical or ellipsoidal	5–70 nm	Silver	Antibacterial	[50]
Erythrophleum fordii	*Alternaria tenuissima*	Spherical	15–45 nm.	Zinc oxide	Antimicrobial, anticancer and antioxidant	[51]
Chonemorpha fragrans.	*Fusarium solani*	Spindle	40–45 nm	Gold	Anticancer	[52]
Cinnamomum zeylanicum	*Lasiodiplodia theobromae*	Spherical to oval	~ 76 nm	Silver	Antibacterial	[53]
Chiliadenus montanus	*Trichoderma atroviride*	Spherical	10 to 15 nm.	Silver	Antibacterial	[54]
Madhuca longifolia	*Pestalotia sp.*	Angular	< 40 nm	Silver	Antibacterial	[55]
Pinus densiflora	*Talaromyces purpureogenus*	Round to triangle	~ 25 nm	Silver	Antimicrobial and anticancer	[56]
Ocimum tenuiflorum	*Exserohilum rostrata,*	Spherical	10–15 nm	Silver	Antibacterial, antiinflammatory, and antioxidant	[57]
Borszczowia aralocaspica	*Isoptericola* SYSU 333150	Spherical	11–40 nm	Silver	Antibacterial	[58]
Oxalis corniculata	*Streptomyces zaomyceticus Oc-5*	Spherical	~78 nm	Copper	Antimicrobial, antioxidant and anticancer	[59]
Mentha longifolia	*Streptomyces sp.*	Spherical	2.3–85 nm	Silver	Antimicrobial	[60]

Plants	Endophytes	Shapes	Size	Types of NPs	Activity	References
Convolvulus arvensis	*Streptomyces capillispiralis* Ca-1,	Spherical	3.6–59 nm	Copper	Antimicrobial and insecticides	[61]
Ocimum sanctum	*Streptomyces coelicolor*	Spherical and ellipsoidal	~25 nm	*Magnesium*	Antimicrobial	[62]

Table 1.
Biosynthesis of nanoparticles from endophytic microorganisms with their respective size and biological activity.

Clarance et al. [52], reported the isolation of the endophytic fungi, *Fusarium solani*, from the plant *Chonemorpha fragrans*, which is used for the biosynthesis of gold nanoparticles. The morphology of synthesized nanomaterials was found to have needled and flower like structures with spindle shape, and showed pink-ruby red colors and high peak plasmon band between 510 and 560 nm. The gold synthesized nanoparticles showed cytotoxic activity against cervical cancer cells (HeLa) (IC50: 0.8 ± 0.5 μg/mL) and human breast cancer cells (MCF-7) (IC50: 1.3 ± 0.5 μg/mL).

Abdelhakim et al. [51], use the culture filtrate of the endophytic fungi *Alternaria tenuissima,* isolated from *Erythrophleum fordii,* to produce zinc oxide nanoparticles. The shape of the biosynthesized nanoparticles was spherical and having size diameter ranges between 15 and 45 nm along with significant antimicrobial, anticancer and antioxidant activity.

The endophyte *Exserohilum rostrata* has been isolated from the plant *Ocimum tenuiflorum* by Bagur et al. [57], this strain was used for the biosynthesis of spherical silver nanoparticles with a size, range between 10 and 15 nm, and showed significant antimicrobial activity and other biological properties such as, anti-inflammatory, and antioxidant activities.

5.3 Nanoparticles synthesized by endophytic actinomycetes

Actinomycetes are Gram positive bacteria with high G + C, belong to the phylum of *Actinobacteria*, which is one of the largest taxonomic rank within the domain of *Bacteria* [68, 69]. This group of microorganisms is known by the production of a wide range of bioactive secondary metabolites. In fact, 70–80% of secondary metabolites in current clinical use, including, antibiotics, antifungals, immunosuppressives, anticancer, insecticides and antivirals, have been isolated and characterized from several species of actinomycetes, particularly from the genus *Streptomyces* [70].

Nanoparticles from endophytic actinobacteria is an emerging field yet to be established, in fact, when compared with fungi and the other bacteria, only few publications have been reported. Most of the articles about nanoparticles from endophytic actinomycetes, reporting the synthesis of nanoparticles using endophytes belong to the genus of *Streptomyces*, however, nanoparticles synthesized by rare actinobacteria have been reported in a few papers [60, 62].

The author, Hassan et al. [59, 61], publishes two papers about the utilization of endophytic *Streptomyces* for the biosynthesis of nanoparticles. In fact, they report the isolation of *Streptomyces zaomyceticus* Oc-5 and *Streptomyces capillispiralis* Ca-1, from the plants *Oxalis corniculata* and *Convolvulus arvensis* respectively. Both strains were used for the synthesis of copper naoparticles, which exhibited different biological activity, including, antimicrobial, antioxidant and anticancer, and insecticides.

In another study, Dong et al. [58], use a rare actinobacteria, in order to control the disease caused by *Staphylococcus warneri* which have a significant impact on human health. The researchers use the strain, *Isoptericola SYSU 333150*, isolated from the plant *Borszczowia aralocaspica*, for the biosynthesis of silver nanoparticle using photo-irradiation with sunlight exposition for different periods, they obtained spherical nanoparticles with a size range between, 11–40 nm, which exhibit antimicrobial activity against the pathogen *S. warneri.*

Several others studies confirm that nanoparticles from different metallic natures, sizes and shapes, synthesized by endophytic microorganisms, are attractive options, since they exhibited various pool of biological activities, including, antimicrobial, cytotoxic, antiinflamatory, antioxidant [35, 50, 53–56, 60, 62].

6. Methods for the isolation of endophytic microorganism and the characterization of synthesized nanoparticles

The isolation methods of endophyte aim to obtained microorganisms reside within plant hosts without causing disease symptoms. The isolation protocol followed depend on several factors such as, the target group of endophyte microorganisms you would like to isolate (bacteria, fungi and Actinobacteria), specie of the host plant, the part of plant tissue, sampling season, culture conditions, etc. [71].

The first step consists on surface sterilization of host plant to remove all the surface-living microorganisms [72]. Several methods can be applied, among them, the plant parts will be immersed sequentially, in several solutions of sterilization, including, 70% ethanol for 5 minutes, followed by (3–10%) of sodium hypochlorite for 2 minutes, and then immersed in hydrogen peroxide (H2O2) for 1 minutes [73]. The final step of sterilization consists to rinse the different plant parts with distilled water three times, and soaked in 10% NaHCO3 to inhibit fungal growth [74].

After surface sterilization, the sterilized tissue samples are cut into small pieces of 1 cm^3, under sterile conditions, and then placed on tryptic soy agar plates followed by incubation for 14 days to verify the sterilization effectiveness. Afterwards, the plant segments are grinding in sterile conditions, and then the samples are serially diluted up to 10^{-3} with sterile water [75]. Aliquots of 100–200 μL of the dilutions will be spread-plated onto a series of appropriate isolation media (depend on the type of endophytic microorganisms). The appeared colonies are transferred to a new culture medium to obtain a pure culture [76]. The endophytic strains are subjected to molecular identification based on sequencing of 16 s rDNA for bacteria, and 18 s r DNA for fungi.

For nanoparticles synthesis, the endophytic strains are culturing in rotating shaker under optimum culture conditions, including: appropriate culture medium, pH, temperature, agitation. After incubation, the culture is centrifuged to separate the biomass from the supernatant [48]. Both supernanant and biomass are tested for nanoparticles synthesize, in fact microorganisms are able to synthesize nanoparticles extracellularly or intracellularly (**Figure 3**) [77].

For extracellular synthesis of nanoparticles, the obtained supernatant is mixed with a filter-sterilized metal salt solution (e.g. AgNo3), the melange is incubated again, the color changing, of the melange after incubation, can indicate the synthesis of nanoparticles [78]. For example, for silver nanoparticles, the color changes from colorless to deep brown, whereas, for gold nanoparticles, it changes from ruby red to a deep purple color. Afterward, the precipitate of nanoparticles formed can be recovered by centrifugation, washed several times with distiled water and collected in the form of a bottom pellet [79].

For intracellular synthesis of nanoparticles, the biomass obtained after centrifugation, is washed several times with distilled water to remove the traces of culture medium, then mixed with a filter-sterilized solution of metal salt [80]. The synthesis of nanoparticles can be monitored by color change after the incubation period [81]. The naoparticules synthesized inside the cell can be released after break down the cell wall by repeated cycles of ultrasonication. The nanoparticles can be purified from cellular debris, after repeated cycles of centrifugation/washing with distilled water [82].

Physicochemical characterization of nanoparticles is performed to determine the morphology, surface area, porosity, particle size and distribution, aggregation, crystal structure (crystallinity), zeta potential, structural properties and others parameters of biosynthetized nanoparticles [40].

In order to analyse the physicochemical properties of nanoparticles, different characterization techniques are applied. This includes the following:

- The formation of nanoparticles can be confirmed by spectra analysis of absorption in the wavelength range between 200 and 800 nm [83].

- The morphology, size and distribution of nanoparticles can be determined by Transmission Electron Microscopy (TEM), as well as Scanning Electron Microscopy (SEM), since morphological features significantly affect the activity of nanoparticles [84].

- The X-ray diffraction (XRD), can be used for the determination of the structural properties of nanoparticles, such as the chemical composition and the crystallinity of synthesized nanoparticles [85].

- FTIR (Fourier transform infrared) spectroscopy, is performed to identify the functional groups present on nanoparticles [86].

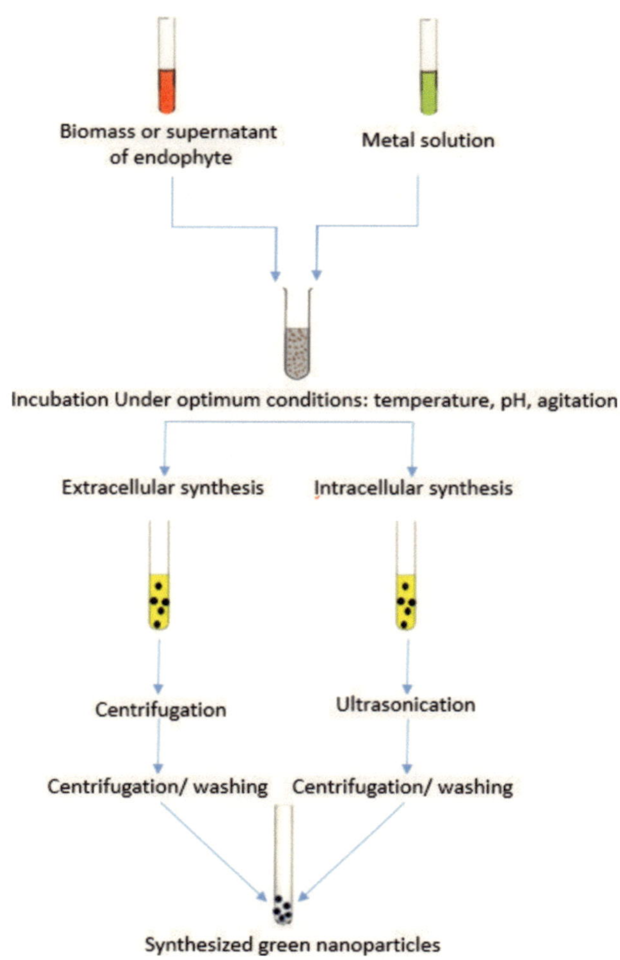

Figure 3.
Green synthesis of nanoparticles using endophyte microorganisms.

- Particle size can be estimated using, dynamic light scattering (DLS), which can be used to find the nanoparticles size at extremely low level [85].
- Surface area characterization, including, stability and surface charge of colloidal nanoparticles are evaluated by zeta potential analysis using a Zetasizer nanomachine [87].

7. Conclusion

Soil microorganisms have been largely explored as a source for nanoparticle biosynthesis; however, few reports are available about the utilization of endophytic microorganisms for synthesizing nanoparticles, and therefore, it is important to focus research in this promising biological route of nanoscience. However, since most of the endophytic microorganisms are uncultivated, it's important to concentrate researchs in the development of innovating methods for the isolation of this group of microorganisms for further advancement of green synthesis of metal nanomaterials. Additionally, the mechanisms involved in the reduction and stabilization of nanoparticles, using microorganisms, is not well defined, and more elaborated studies are needed to determine all the enzymes and biomolecules involved in the nanoparticle biosynthesis.

Author details

Omar Messaoudi[1,2]* and Mourad Bendahou[2]

1 Department of Biology, Faculty of Science, University of Amar Telidji, Laghouat, Algeria

2 Laboratory of Applied Microbiology in Food and Environment, Abou bekr Belkaïd University, Tlemcen, Algeria

*Address all correspondence to: o.messaoudi@lagh-univ.dz

References

[1] Porter AL, Youtie J. How interdisciplinary is nanotechnology? Journal of Nanoparticle Research. 2009;**11**(5):1023-1041. DOI: 10.1007/s11051-009-9607-0

[2] Khan I, Saeed K, Khan I. Nanoparticles: Properties, applications and toxicities. Arabian Journal of Chemistry. 2019;**12**(7):908-931. DOI: 10.1016/j.arabjc.2017.05.011

[3] Malik B, Pirzadah TB, Kumar M, Rehman RU. Biosynthesis of nanoparticles and their application in pharmaceutical industry. In: Prasad R, Kumar V, Kumar M, editors. Nanotechnology. Singapore: Springer; 2017. pp. 235-252. DOI: 10.1007/978-981-10-4678-0_13

[4] Iravani S. Bacteria in nanoparticle synthesis: Current status and future prospects. International Scholarly Research Notices. 2014;**2014**:1-18. DOI: 10.1155/2014/359316

[5] Singh J, Dutta T, Kim KH, Rawat M, Samddar P, Kumar P. 'Green' synthesis of metals and their oxide nanoparticles: Applications for environmental remediation. Journal of Nanbiotechnology. 2018;**16**:84. DOI: 10.1186/s12951-018-0408-4

[6] Gour A, Jain NK. Advances in green synthesis of nanoparticles. Artificial Cells, Nanomedicine, and Biotechnology. 2019;**47**:844-851. DOI: 10.1080/21691401.2019.1577878

[7] Grasso G, Zane D, Dragone R. Microbial nanotechnology: Challenges and prospects for green biocatalytic synthesis of nanoscale materials for sensoristic and biomedical applications. Nanomaterials. 2020;**10**:11. DOI: 10.3390/nano10010011

[8] Li X, Xu H, Chen ZS, Chen G. Biosynthesis of nanoparticles by microorganisms and their applications. Journal of Nanomaterials. 2011;**2011**:270974. DOI: 10.1155/2011/270974

[9] Yadav KK, Singh JK, Gupta N, Kumar V. A review of nanobioremediation technologies for environmental cleanup: A novel biological approach. Journal of Materials and Environmental Science. 2017;**8**:740-757

[10] Rana KL, Kour D, Yadav N, Yadav AN. Endophytic microbes in nanotechnology: Current development, and potential biotechnology applications. In: Kumar A, Singh VK, editors. Microbial Endophytes. Cambridge, MA: Woodhead Publishing; 2020. pp. 231-262. DOI: 10.1016/B978-0-12-818734-0.00010-3

[11] Nath R, Sharma G, Barooah M. Plant growth promoting endophytic fungi isolated from tea (Camellia sinensis) shrubs of Assam. Applied Ecology and Environmental Research. 2015;**13**:877-891. DOI: 10.15666/aeer/1303_877891

[12] Glassner H, Zchori-Fein E, Yaron S, Sessitsch A, Sauer U, Compant S. Bacterial niches inside seeds of Cucumis melo L. Plant and Soil. 2017;**422**:1-3. DOI: 10.1007/s11104-017-3175-3

[13] Strobel G, Daisy B. Bioprospecting for microbial endophytes and their natural products. Microbiology and Molecular Biology Reviews. 2003;**67**:491-502. DOI: 10.1128/MMBR.67.4.491-502.2003

[14] Hardoim PR, Van Overbeek LS, Berg G, Pirttilä AM, Compant S, Campisano A, et al. The hidden world within plants: Ecological and evolutionary considerations for defining functioning of microbial endophytes.

Microbiology and Molecular Biology Reviews. 2015;**79**:293-320. DOI: 10.1128/MMBR.00050-14

[15] Ahemad M, Kibret M. Mechanisms and applications of plant growth promoting rhizobacteria: Current perspective. Journal of King Saud University – Science. 2014;**26**:1-20. DOI: 10.1016/j.jksus.2013.05.001

[16] Verma A, Kukreja K, Pathak DV, Suneja S, Narula N. In vitro production of plant growth regulators (PGRs) by Azotobacter chroococcum. Indian Journal of Microbiology. 2001;**41**:305-307

[17] Antoun H, Prévost D. Ecology of plant growth promoting rhizobacteria. In: Siddiqui ZA, editor. PGPR: Biocontrol and Biofertilization. Netherlands: Springer; 2005. pp. 1-38. DOI: 10.1007/1-4020-4152-7_1

[18] Truyens S, Weyens N, Cuypers A, Vangronsveld J. Bacterial seed endophytes: Genera, vertical transmission and interaction with plants. Environmental Microbiology Reports. 2014;**7**:40-50. DOI: 10.1111/1758-2229.12181

[19] Frank A, Saldierna-Guzmán J, Shay J. Transmission of bacterial endophytes. Microorganism. 2017;**5**:E70. DOI: 10.3390/microorganisms5040070

[20] Monowar T, Rahman MS, Bhore SJ, Raju G, Sathasivam KV. Silver nanoparticles synthesized by using the endophytic bacterium Pantoea ananatis are promising antimicrobial agents against multidrug resistant bacteria. Molecules. 2018;**23**:3220. DOI: 10.3390/molecules23123220

[21] Afzal I, Shinwari ZK, Sikandar S, Shahzad S. Plant beneficial endophytic bacteria: Mechanisms, diversity, host range and genetic determinants. Microbiological Research. 2019;**221**:36-49. DOI: 10.1016/j.micres.2019.02.001

[22] Akinsanya MA, Goh JK, Lim SP, Ting AY. Metagenomics study of endophytic bacteria in Aloe vera using next-generation technology. Genomics Data. 2015;**6**:159-163. DOI: 10.1016/j.gdata.2015.09.004

[23] Vieira ML, Johann S, Hughes FM, Rosa CA, Rosa LH. The diversity and antimicrobial activity of endophytic fungi associated with medicinal plant Baccharis trimera (Asteraceae) from the Brazilian savannah. Canadian Journal of Microbiology. 2014;**60**:847-856. DOI: 10.1139/cjm-2014-0449

[24] Li JL, Sun X, Zheng Y, Lü PP, Wang YL, Guo LD. Diversity and community of culturable endophytic fungi from stems and roots of desert halophytes in Northwest China. MycoKeys. 2020;**62**:75-95. DOI: 10.3897/mycokeys.62.38923

[25] Teimoori-Boghsani Y, Ganjeali A, Cernava T, Müller H, Asili J, Berg G. Endophytic fungi of native salvia abrotanoides plants reveal high taxonomic diversity and unique profiles of secondary metabolites. Frontiers in Microbiology. 2020;**10**:3013. DOI: 10.3389/fmicb.2019.03013

[26] Abdel-Aziz SM, Prasad R, Hamed AA, Abdelraof M. Fungal nanoparticles: A novel tool for a green biotechnology? In: Prasad R, Kumar V, Kumar M, Wang S, editors. Fungal Nanobionics: Principles and Applications. Singapore: Springer; 2018. pp. 61-87. DOI: 10.1007/978-981-10-8666-3_3

[27] Valavanidis A, Lachogiann T. Engineered nanomaterials for pharmaceutical and biomedical products new trends, benefits and opportunities. Pharmaceutical Bioprocessing. 2016;**4**(1):13-24

[28] Wang Y, Xia Y. Bottom-up and top-down approaches to the synthesis of monodispersed spherical colloids of low melting-point metals. Nano Letters. 2004;(4):2047-2050. DOI: 10.1021/nl048689j

[29] Priyadarshana G, Kottegoda N, Senaratne A, Karunaratne V. Synthesis of magnetite nanoparticles by top-down approach from a high purity ore. Journal of Nanomaterials. 2015;**2015**:1-8. DOI: 10.1155/2015/317312

[30] Merkel TJ, Herlihy KP, Nunes J, Orgel RM, Rolland JP, DeSimone JM. Scalable, shape-specific, top-down fabrication methods for the synthesis of engineered colloidal particles. Langmuir. 2010;**26**(16):13086-13096. DOI: 10.1021/la903890h

[31] Daraio C, Jin S. Synthesis and patterning methods for nanostructures useful for biological applications. In: Silva G, Parpura V, editors. Nanotechnology for Biology and Medicine. Fundamental Biomedical Technologies. New York: Springer; 2012. pp. 27-44. DOI: 10.1007/978-0-387-31296-5_2

[32] Kiranmai M. Biological and non-biological synthesis of metallic nanoparticles: Scope for current pharmaceutical research. Indian Journal of Pharmaceutical Sciences. 2017;**79**:501-512. DOI: 10.4172/pharmaceutical-sciences.1000256

[33] Saranya S, Vijayarani K, Pavithra S. Green synthesis of iron nanoparticles using aqueous extract of musa ornate flower sheath against pathogenic bacteria. Indian Journal of Pharmaceutical Sciences. 2017;**79**:688-694. DOI: 10.4172/pharmaceutical-sciences.1000280

[34] Patra JK, Baek KH. Green nanobiotechnology: Factors affecting synthesis and characterization techniques. Journal of Nanomaterials. 2014;**2014**:219. DOI: 10.1155/2014/417305

[35] Singh T, Jyoti K, Patnaik A, Singh A, Chauhan R, Chandel SS. Biosynthesis, characterization and antibacterial activity of silver nanoparticles using an endophytic fungal supernatant of Raphanus sativus. Journal, Genetic Engineering & Biotechnology. 2017;**15**:31-39. DOI: 10.1016/j.jgeb.2017.04.005

[36] Al-Sheddi ES, Farshori NN, Al-Oqail MM, Al-Massarani SM, Saquib Q, Wahab R, et al. Anticancer potential of green synthesized silver nanoparticles using extract of Nepeta deflersiana against human cervical cancer cells (HeLA). Bioinorganic Chemistry and Applications. 2018;**2018**:9390784. DOI: 10.1155/2018/9390784

[37] Soliman H, Elsayed A, Dyaa A. Antimicrobial activity of silver nanoparticles biosynthesised by Rhodotorula sp. strain ATL72. Egyptian Journal of Basic and Applied Sciences. 2018;**5**(3):228-233. DOI: 10.1016/j.ejbas.2018.05.005

[38] Khandel P, Yadaw RK, Soni DK, Kanwar L, Shahi SK. Biogenesis of metal nanoparticles and their pharmacological applications: Present status and application prospects. Journal of Nanostructure in Chemistry. 2018;**8**:217-254. DOI: 10.1007/s40097-018-0267-4

[39] Wang JC, Neogi P, Forciniti D. On one-dimensional self-assembly of surfactant-coated nanoparticles. The Journal of Chemical Physics. 2006;**125**(19):194717. DOI: 10.1063/1.2375091

[40] Ghashghaei S, Emtiazi G. The methods of nanoparticle synthesis using bacteria as biological nanofactories, their mechanisms and major applications. Current

Bionanotechnology. 2015;**1**:3-17. DOI: 10.2174/2213529401999140310104655

[41] Talekar S, Ghodake V, Ghotage T, Rathod P, Deshmukh P, Nadar S, et al. Novel magnetic cross-linked enzyme aggregates (magnetic CLEAs) of alpha amylase. Bioresource Technology. 2012;**123**:542-547. DOI: 10.1016/j.nanoso.2016.03.002

[42] Velusamy P, Kumar GV, Jeyanthi V, Das J, Pachaiappan R. Bio-inspired green nanoparticles: Synthesis, mechanism, and antibacterial application. Toxicology Research. 2016;**32**(2):95-102. DOI: 10.5487/TR.2016.32.2.095

[43] Chokkareddy R, Redhi GG. Green synthesis of metal nanoparticles and its reaction mechanisms. In: Kanchi S, Ahmed S, editors. Green Metal Nanoparticles: Synthesis, Characterization and Their Application. Beverly, MA, USA: Scrivener Publishing LLC; 2018. pp. 113-139. DOI: 10.1002/9781119418900.ch4

[44] Niño-Martínez N, Salas Orozco MF, Martínez-Castañón GA, Torres Méndez F, Ruiz F. Molecular mechanisms of bacterial resistance to metal and metal oxide nanoparticles. International Journal of Molecular Sciences. 2019;**20**:2808-2823. DOI: 10.3390/ijms20112808

[45] Joshi CG, Danagoudar A, Poyya J, Kudva AK, Dhananjaya B. Biogenic synthesis of gold nanoparticles by marine endophytic fungus-Cladosporium cladosporioides isolated from seaweed and evaluation of their antioxidant and antimicrobial properties. Process Biochemistry. 2017;**63**:137-144. DOI: 10.1016/j.procbio.2017.09.008

[46] Ibrahim E, Fouad H, Zhang M, Zhang Y, Qiu W, Yan C, et al. Biosynthesis of silver nanoparticles using endophytic bacteria and their role in inhibition of rice pathogenic bacteria and plant growth promotion. RSC Advances. 2019;**9**:29293-29299. DOI: 10.1039/C9RA04246F

[47] Ibrahim E, Zhang M, Zhang Y, Hossain A, Qiu W, Chen Y, et al. Green-synthesization of silver nanoparticles using endophytic bacteria isolated from garlic and its antifungal activity against wheat Fusarium head blight pathogen Fusarium graminearum. Nanomaterials. 2020;**10**:219. DOI: 10.3390/nano10020219

[48] Prathna TC, Lazar Mathew NC, Ashok M, Raichur AM. Biomimetic synthesis of nanoparticles: Science, technology & amp. In: Amitava M, editor. Applicability, Biomimetics Learning from Nature. Rijeka: IntechOpen; 2010. DOI: 10.5772/8776

[49] Syed B, Prasad NM, Satish S. Endogenic mediated synthesis of gold nanoparticles bearing bactericidal activity. Journal of Microscopy and Ultrastructure. 2016;**4**:162-166. DOI: 10.1016/j.jmau.2016.01.004

[50] Farsi M, Farokhi S. Biosynthesis of antibacterial silver nanoparticles by endophytic fungus Nemania sp. isolated from Taxus baccata L. (Iranian yew). Zahedan Journal of Research in Medical Sciences. 2018;**20**(6):e57916. DOI: 10.5812/zjrms.57916

[51] Abdelhakim HK, El-Sayed ER, Rashidi FB. Biosynthesis of zinc oxide nanoparticles with antimicrobial, anticancer, antioxidant and photocatalytic activities by the endophytic Alternaria tenuissima. Journal of Applied Microbiology. 2020;**128**:1634-1646. DOI: 10.1111/jam.14581

[52] Clarance P, Luvankar B, Sales J, Khusro A, Agastian P, Tack JC, et al. Green synthesis and characterization of gold nanoparticles using endophytic fungi Fusarium solani and its invitro

anticancer and biomedical applications. Saudi Journal of Biological Sciences. 2020;**27**:706-712. DOI: 10.1016/j.sjbs.2019.12.026

[53] Ranjani S, Shariq AM, Mohd A, Senthil KN, Ruckmani K, Hemalatha S. Synthesis, characterization and applications of endophytic fungal nanoparticles. Inorganic and Nano-Metal Chemistry. 2020;**27**(2):706-712. DOI: 10.1080/24701556.2020.178423

[54] Abdel-Azeem A, Nada AA, O'Donovan A, Kumar TV, Elkelish A. Mycogenic silver nanoparticles from endophytic trichoderma atroviride with antimicrobial activity. Journal of Renewable Materials. 2019;7:171-185. DOI: 10.32604/jrm.2020.08960

[55] Verma SK, Gond SK, Mishra A, Sharma VK, Kumar J, et al. Biofabrication of antibacterial and antioxidant silver nanoparticles (Agnps) by an endophytic fungus Pestalotia Sp. isolated from Madhuca Longifolia. Journal of Nanomaterials & Molecular Nanotechnology. 2016;**5**:3. DOI: 10.4172/2324-8777.1000189

[56] Xiaowen H, Saravankumar K, Jin T, Myeong-Hyeon W. Mycosynthesis, characterization, anticancer and antibacterial activity of silver nanoparticles from endophytic fungus Talaromyces purpureogenus. International Journal of Nanomedicine. 2019;**14**:3427-3438. DOI: 10.2147/IJN.S200817

[57] Bagur H, Poojari CC, Melappa G, et al. Biogenically synthesized silver nanoparticles using endophyte fungal extract of Ocimum tenuiflorum and evaluation of biomedical properties. Journal of Cluster Science. 2019;**30**:1-15. DOI: 10.1007/s10876-019-01731-4

[58] Dong ZY, Narsing Rao MP, Xiao M, Wang HF, Hozzein WN, Chen W, et al. Antibacterial activity of silver nanoparticles against Staphylococcus warneri synthesized using endophytic bacteria by photo-irradiation. Frontiers in Microbiology. 2017;**8**:1090. DOI: 10.3389/fmicb.2017.01090

[59] Hassan SE-D, Fouda A, Radwan AA, Salem SS, Barghoth MG, Awad MA, et al. Endophytic actinomycetes Streptomyces spp mediated biosynthesis of copper oxide nanoparticles as a promising tool for biotechnological applications. Journal of Biological Inorganic Chemistry. 2019;**24**(3):377-393. DOI: 10.1007/s00775-019-01654-5

[60] El-Gamal MS, Salah SS, Salem S, Abdullah A. Biosynthesis, characterization, and antimicrobial activity of silver nanoparticles synthesized by endophytic Streptomyces sp. Egyptian Journal of Biotechnology. 2018;**56**:69-85

[61] Hassan SELD, Salem SS, Fouda A, Awad MA, El-Gamal MS, Abdo AM. New approach for antimicrobial activity and bio-control of various pathogens by biosynthesized copper nanoparticles using endophytic actinomycetes. Journal of Radiation Research and Applied Science. 2018;**11**(3):262-270. DOI: 10.1016/j.jrras.2018.05.003

[62] El-Moslamy SH. Bioprocessing strategies for cost-effective large-scale biogenic synthesis of nano-MgO from endophytic Streptomyces coelicolor strain E72 as an anti-multidrug-resistant pathogens agent. Scientific Reports. 2018;**8**(1):3820. DOI: 10.1038/s41598-018-22134-x

[63] Iravani S, Korbekandi H, Mirmohammadi SV, Zolfaghari B. Synthesis of silver nanoparticles: Chemical, physical and biological methods. Research in Pharmaceutical Sciences. 2014;**9**(6):385-406

[64] Syed B, Prasad MN, Satish S. Synthesis and characterization of silver nanobactericides produced by

Aneurinibacillus migulanus 141, a novel endophyte inhabiting Mimosa pudica L. Arabian Journal of Chemistry. 2019;**12**(8):3743-3752. DOI: 10.1016/j.arabjc.2016.01.005

[65] Moghaddam AB, Namvar F, Moniri M, Tahir PM, Azizi S, Mohamad R. Nanoparticles biosynthesized by fungi and yeast: A review of their preparation, properties, and medical applications. Molecules. 2015;**20**(9):16540-16565. DOI: 10.3390/molecules200916540

[66] Shah M, Fawcett D, Sharma S, Tripathy SK, Poinern GEJ. Green of metallic synthesis nanoparticles via biological entities. Material. 2015;**8**:7278-7308. DOI: 10.3390/ma8115377

[67] Guilger-Casagrande M, Lima R. Synthesis of silver nanoparticles mediated by fungi: A review. Frontiers in Bioengineering and Biotechnology. 2019;**22**(7):287. DOI: 10.3389/fbioe.2019.00287

[68] Ait Barka EA, Vatsa P, Sanchez L, Gaveau-Vaillant N, Jacquard C, Klenk HP, et al. Taxonomy, physiology, and natural products of Actinobacteria. Microbiology and Molecular Biology Reviews. 2016;**80**:1-43. DOI: 10.1128/MMBR.00019-15

[69] Messaoudi O, Sudarman E, Bendahou M, Jansen R, Stadler M, Wink J. Kenalactams A–E, polyene macrolactams isolated from nocardiopsis CG3. Journal of Natural Products. 2019;**82**:1081-1088. DOI: 10.1021/acs.jnatprod.8b00708

[70] Messaoudi O, Bendahou M, Benamar I, Abdelwouhid DE. Identification and preliminary characterization of non-polyene antibiotics secreted by new strain of actinomycete isolated from sebkha of Kenadsa, Algeria. Asian Pacific Journal of Tropical Biomedicine. 2015;**5**:438-445. DOI: 10.1016/j.apjtb.2015.04.002

[71] Lu Y, Chen C, Chen H, Zhang J, Chen W. Isolation and identification of endophytic fungi from Actinidia macrosperma and investigation of their bioactivities. Evidence-based Complementary and Alternative Medicine. 2012;**2012**:382742. DOI: 10.1155/2012/382742

[72] Passari AK, Misha VK, Gupta VK, Singh BP. Chapter 1: Methods used for the recovery of culturable endophytic actinobacteria: An overview. In: Singh BP, Passari AK, Gupta VK, editors. Actinobacteria: Diversity and Biotechnological Applications: New and Future Developments in Microbial Biotechnology and Bioengineering. Amsterdam: Elsevier; 2018. pp. 1-11

[73] Hallmann J, Berg G, Schulz B. Isolation procedures for endophytic microorganisms. In: Schulz BJE, Boyle CJC, Sieber TN, editors. Microbial Root Endophytes. New York: Springer; 2006. pp. 299-314

[74] Nimnoi P, Pongsilp N, Lumyong S. Endophytic actinomycetes isolated from Aquilaria crassna Pierre ex Lec and screening of plant growth promoters production. World Journal of Microbiology and Biotechnology. 2010;**26**:193-203. DOI: 10.1007/s11274-009-0159-3

[75] Taechowisan T, Lumyong S. Activity of endophytic actinomycetes from roots of Zingiber officinale and Alpinia galena against phytopathogenic fungi. Annales de Microbiologie. 2003;**53**:291-298

[76] Passari AK, Mishra VK, Saikia R, Gupta VK, Singh BP. Isolation, abundance and phylogenetic affiliation of endophytic actinomycetes associated with medicinal plants and screening for their in vitro antimicrobial biosynthetic potential. Frontiers in

Microbiology. 2015;**6**:273. DOI: 10.3389/fmicb.2015.00273

[77] Vetchinkina E, Loshchinina E, Kupryashina M, Burov A, Pylaev T, Nikitina V. Green synthesis of nanoparticles with extracellular and intracellular extracts of basidiomycetes. PeerJ. 2018;**6**:e5237. DOI: 10.7717/peerj.5237

[78] Karthik L, Kumar G, Vishnu-Kirthi A, Rahuman AA, Rao VB. Streptomyces sp. LK3 mediated synthesis of silver nanoparticles and its biomedical application. Bioprocess and Biosystems Engineering. 2014;**37**:261-267. DOI: 10.1007/s00449-013-0994-3

[79] Singh H, Du J, Singh P, Yi TH. Extracellular synthesis of silver nanoparticles by pseudomonas sp. THG-LS1.4 and their antimicrobial application. Journal of Pharmaceutical Analysis. 2018;**8**:258-264. DOI: 10.1016/j.jpha.2018.04.004

[80] Ahmad A, Mukherjee P, Senapati S, Mandal D, Khan M, Kumar R, et al. Extracellular biosynthesis of silver nanoparticles using the fungus Fusarium oxysporum. Colloids and Surfaces, B: Biointerfaces. 2003;**28**:313-318. DOI: 10.1002/cbic.200700592

[81] Abdeen S, Geo S, Sukanya S, Praseetha PK, Dhanya RP. Biosynthesis of silver nanoparticles from Actinomycetes for therapeutic applications. International Journal of Nano Dimension. 2014;**5**:155-162. DOI: 10.7508/IJND.2014.02.008

[82] Malik P, Shankar R, Malik V, Sharma N, Mukherjee TK. Green chemistry based benign routes for nanoparticle synthesis. Journal of Nanoparticle Research. 2014;**2014**:302429. DOI: 10.1155/2014/302429

[83] Ingale AG, Chaudhari NA. Biogenic synthesis of nanoparticles and potential applications: An eco-friendly approach. Nanomaterials, Nanotechnology and Nanomedicine. 2013;**4**(2):165. DOI: 10.4172/2157-7439.1000165

[84] Pal SL, Jana U, Manna PK, Mohanta GP, Manavalan R. Nanoparticle: An overview of preparation and characterization. Journal of Applied Pharmacology. 2011;**1**(6):228-234

[85] Chauhan RPS, Gupta C, Prakash D. Methodological advancements in green nanotechnology and their applications in biological synthesis of herbal nanoparticles. International Journal of Bioassays. 2012;**1**(7):6-10

[86] Faraji M, Yamini Y, Rezaee M. Magnetic nanoparticles: Synthesis, stabilization, functionalization, characterization, and applications. Journal of the Iranian Chemical Society. 2010;**7**(1):1-37. DOI: 10.1007/BF03245856

[87] Otsuka H, Nagasaki Y, Kataoka K. PEGylated nanoparticles for biological and pharmaceutical applications. Advanced Drug Delivery Reviews. 2003;**55**(3):403-419

Section 2

Nanomaterials in Energy Storage

Chapter 4

Nanotechnology in the Service of Solar Energy Systems

Farzaneh Ghasemzadeh and Mostafa Esmaeili Shayan

Abstract

Nanotechnology can help to address the existing efficiency hurdles and greatly increase the generation and storage of solar energy. A variety of physical processes have been established at the nanoscale that can improve the processing and transmission of solar energy. The application of nanotechnology in solar cells has opened the path to the development of a new generation of high-performance products. When competition for clean energy options is growing, a variety of potential approaches have been discussed in order to expand the prospects. New principles have been explored in the area of solar cell generation, multi-generation, spectrum modulation, thermo-photoelectric cells, hot carrier, the middle band, and many other techniques. Nanoparticles and nanostructures have been shown to enhance the absorption of light, increase the conversion of light to energy, and have improved thermal storage and transport.

Keywords: nanotechnology, thin-film, solar cells, renewable energy

1. Introduction

Nanotechnology is an interdisciplinary area of research, engineering and development that encompasses nanoscale materials from 1 to 100 nanometers. At these nanometer measurements, materials can exhibit new properties that are absent or low in their bulk. For this purpose, nanotechnology applications have been proved in a broad variety of areas, such as physics, chemistry, biological sciences, materials sciences, electronics and energy sciences. Owing to the dwindling fossil fuel supplies and the environmental consequences of their usage and the rising greenhouse gas emissions that have warmed the earth, discovering a new source of renewable, efficient and biocompatible energy is a problem confronting scientists and researchers today [1]. Attention to alternative energy sources such as solar energy, wind, hydro and tides, and biomass from fuel cells and hydrogen is of great interest in industrial and science communities today [2]. Solar energy, the world's biggest energy source, as a renewable energy source, inexpensive and free emissions, has a special role in energy supply. The sunlight that the Sun reflects on the planet every hour is greater than the entire resources that the people of the world eat in a year [3]. The production of this energy is therefore of considerable significance. Currently, owing to the need to harvest solar energy, numerous forms of solar power production systems have been developed in different countries with the correct capacity for solar radiation, so that their electricity is transmitted to the national grid [4]. Todays, about 178

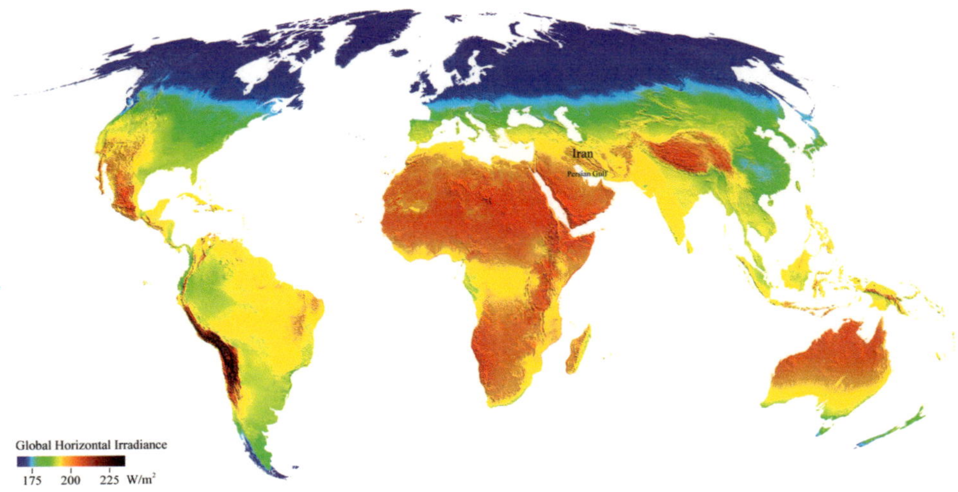

Figure 1.
Global horizontal irradiation.

GW of electricity supply has produced by solar energy [5]. This resource came to expected that, with the elimination of barriers, the solar power potential of built power plants would hit more than 500 GW by 2020 [6]. Since the underlying essence of green energies, like solar energy, is mostly local and distributed, so if it is not feasible to utilize the global grid, it is reasonable that the usage of such resources would be decentralized so spread, rather than centralized output, will be more justifiable. **Figure 1** shows the influence of solar energy in the World. Iran has a strong potential to get this energy supply close to the equator. In the meantime, Iran ranks seventh in mine resources.

It must be remembered that solar energy can be used in two ways [7]:

1. Using sunlight to produce energy directly

2. Usage of solar thermal energy in high-temperature power plants for the generation of electricity and in low-temperature power plants for the processing of hot water and the ventilation of houses, as well as for use of solar water desalination plants

Nanotechnology can be very successful in the extraction of solar energy in this field. Using this technique, the performance of the system can be improved in the two areas of power and heat production.

It is now easier to learn more about the solar power production program and get acquainted with the specifications of this sector.

2. Solar power generation mechanism

Sunlight comprises of various hair lengths of light continuum (ultraviolet, yellow, and red) from energy packets called photons. The intensity of these photons differs based on their wavelength [8]. Sunlight after exposure to the surface of the solar Panels, solar cells absorb the Sun's energy and turn it into electricity (**Figure 2**).

A solar cell is a semiconductor electron that transforms sunlight energy directly to electricity through its photovoltaic influence. If the sunlight passes to the semiconductor, the electron moves from the capacitance band to the semiconductor conduction band and generates an electron–hole pair, each of which can engage in the semiconductor load transfer cycle and create possible variations such that, with the intervention of the user, the load can be guided to the external circuit (**Figure 3**).

Figure 2.
A simple circuit of photovoltaic [7].

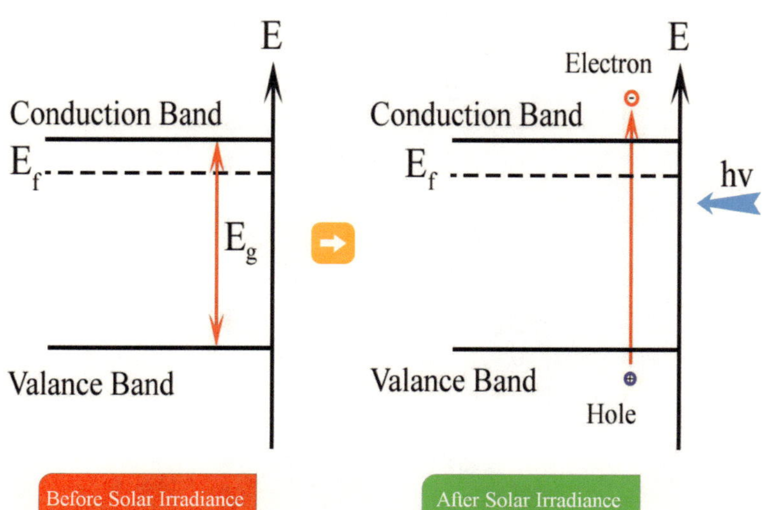

Figure 3.
Schematic of electron and hole in solar irradiation present and non-solar irradiation [9].

3. Types of solar cells

3.1 Silicon solar cells (First generation)

The technology of the first generation is based on silicon wafers with a thickness of between 300 and 400 microns with a single crystal or multi-crystal structure. The silicone materials used are polluted with different elements in order to locate a sufficient amount of electrons-holes. Such solar cells are made up of a combination of electron-contaminated and perforated silicone layers that emit electron-cavity light [7]. Therefore, an electrical current is created by moving the load to the external circuit. These solar cells have been commercialized because of their high performance, but the major drawbacks of this group are the high cost of processing silicon raw materials and the high energy usage [10] (**Figure 4**).

3.2 Thin film of solar cells (Second generation)

As the name suggests, the operating concepts of these cells are focused on thin layers of semiconductors deposited on the surface, such as glass, metal or polymer substrates. In such solar cells, each coating is responsible for consuming part of the Sun's wavelength [11]. As a consequence, the absorption rate in this form of solar cells decreases and their energy transfer improves.

3.3 Dye-sensitized solar cell (Third generation)

The essential components of a pigmented solar cell are a photo-electrode composed of a pigment-sensitive titanium dioxide film (TiO_2). In this cells as Showed in **Figure 5**, when photons are incoming into the Dye-Sensitized Solar Cells, absorbed by the pigment and produced electrons and holes [12]. The electrons in the dye are passed to the nanoparticle of TiO_2. The nanoparticles of TiO_2 serve as carriers of

Figure 4.
Silicon solar panels [7].

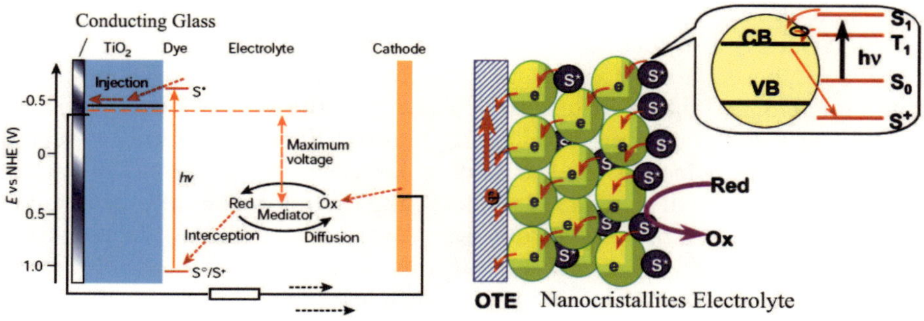

Figure 5.
Schematic of dye-sensitized solar cells [13].

this electron, and finally the electron enters the electrode. The electron that enters the electrode is passed by wire to the opposite electrode (reducing electrode) and is used in the electrolyte recovery cycle. The pigment cavity, on the other side, is regenerated by a solvent and able to absorb the next photon. In this way, the current is also formed on the external circuit.

4. The application of nanotechnology in solar cells industry

Generally, nanotechnology has a beneficial impact on the efficiency of solar cells in a variety of respects [14]. Such functional consequences shall include:

1. Boost absorption and retention of sunlight

2. Providing modern nanotechnology designs for solar cells

3. Using nanowires to boost solar cell efficiency

4. Application of nanotechnology dependent photo-catalysts in solar cells

5. Application of nanocoatings

6. Application of nanotechnology in power storage systems

The list is only several of the features of nanotechnology in the field of solar energy and nanotechnology in other fields of solar energy, such as solar thermal systems, has already demonstrated major applications.

5. The improve of absorption and capture of sunlight

5.1 Light emitting nanoparticles; a reasonable way to improve the performance of a solar cell

Nanomaterials have been developed and manufactured in numerous ways for diverse uses since the advent of nanotechnology in the last decades. Meanwhile, the light emitting nanoparticles got a lot of coverage owing to their very fascinating properties in the area of light absorption and reflection [5]. Light emitting nanoparticles, such as quantum dots, gold or silver nanoparticles and fluorescent nanofibers, are commonly used to improve the performance of solar cells. The basic denominator of such nanoparticles is their special optical properties. Simply stated, the key characteristic of such nanoparticles is the fact that they are fluorescent. Such nanoparticles, based on their shape and scale, may absorb various wavelengths and become agitated, and then release absorbed energy in the form of radiation from another wavelength or original wavelength. Quantum dots will be used as a replacement for pigments owing to their outstanding optoelectronic properties in solar cells [15]. This leads to the likelihood of increasing the voltage or output current of the responsive solar cell to the quantum level. Another drawback of such materials is that they will extend the spectrum of absorption of sunlight outside visible light due to their photoelectric properties (conversion of solar energy into electricity).

Although they also receive infrared rays, quantum-based cell theory simulation has projected an improvement in cell performance of about 64 percent, which is quite important. The light emitting in some of the quantum dots materials are shows in **Figure 6**.

One of the newest products to improve the response of solar cells to light and improve their absorption is the quantum dots of silver sulfide (Ag_2S). Such quantum dots are immune to wavelengths between 400 and 1000 nm in the solar spectrum; thus, in addition to visible light (400–700 nm), they often reflect the intensity of infrared radiation. The genus of this category is silver selenide (Ag_2Se) quantum dots, which span the entire solar spectrum and have a susceptibility range

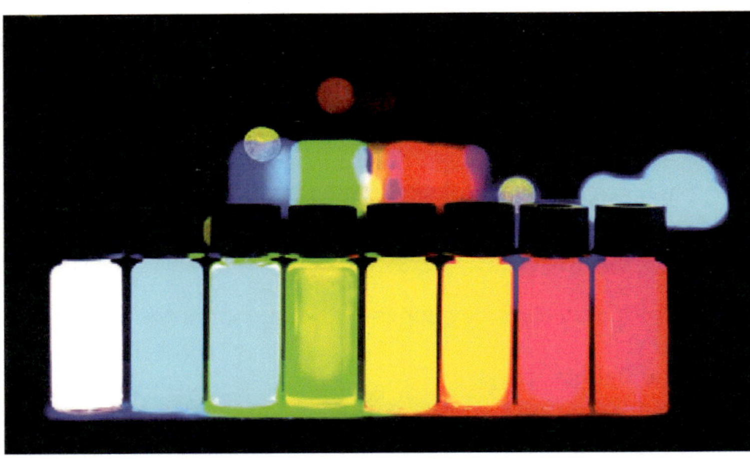

Figure 6.
The light emitting of quantum dots [16].

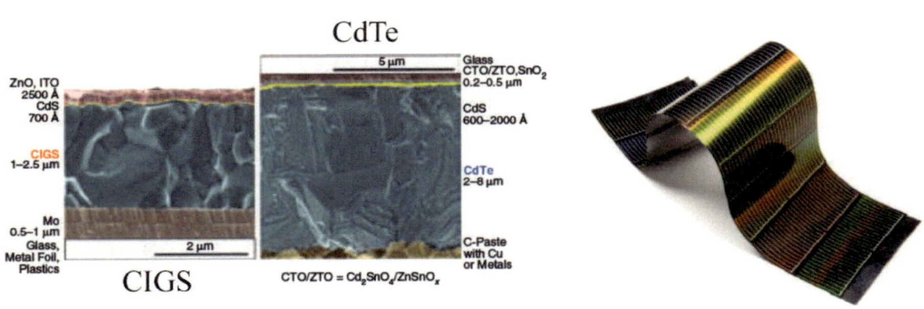

Figure 7.
CIGS and CdTe flexible solar technology [16].

7–14 times that of standard quantum dots [7]. The electrical current production of these cells is 4 times that of regular pigment-sensitive cells. It can also be noted that such quantum dots are very strong choices for increasing the performance of solar cells and, as a result, investing in further work in this area in order to enter the commercial process is completely important and imminent (**Figure 7**).

6. Provide new architectures for solar cells focused on nanotechnology

6.1 Transparent solar cells: A change in the solar industry

With the emergence of nanotechnology in the area of solar cell manufacturing and the development of modern architectures, new opportunities for different forms of solar cell applications have been opened to the world of business and technology. All of these new designs is to render solar cells fully translucent. The ultimate composition of transparent solar cells is a mixture of translucent substrate (made of glass or plastic) and Nanolayers of materials of specific optical properties and thicknesses, which are responsible for absorption beyond the spectrum of visible light. Transparent cells emit visible light and then consume ultraviolet light and infrared-producing electricity. This innovative aspect of translucent solar cells enables a broad variety of applications in buildings and cars. The volume of visible light in various forms of cells ranges from 50% to 80%. Researchers expect that, with the aid of nanotechnology, it would not be difficult to reach 12% performance without compromising the properties of cell movement [17]. The thickest layer of this system is the sheet of glass or plastic to which the sheet of grating and coating is added. Some coatings are added to the ground in nanoscale. At the middle of the layers are two active cells, which absorb stimulated light and emit electrons. One of these two compounds is chloroaluminium phthalocyanine, which serves as an organic electron donor, and the other is carbon 60 (C_{60}) electron receptor [18]. The thickness of phthalocyanine chloroalumine is 15 and C-60 is 30 nm. The electrodes are mounted on all sides of these walls. The electrodes are constructed of ITO/MoO_3. The width of such electrodes is less than 20 nm [19]. As the electrodes may be translucent (and not constructed of ordinary metal), a coating at the end of

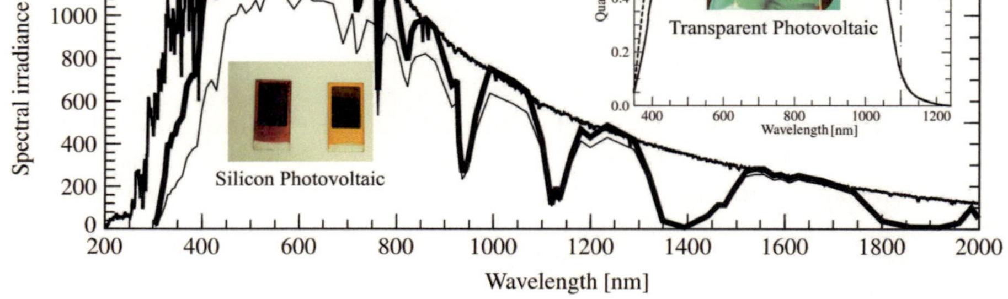

Figure 8.
Effect of spectral and wavelength on transparency photovoltaic and silicon based photovoltaic [20].

the cell can be inserted to replicate the length of the thin from other positions in the sunlight and return it to the cell atmosphere in such a way that the cycle of absorption and conversion is replicated, as shown in **Figure 8**. It is found that, during this special feature, much of the visible light moves across and out. The short and long line extensions in the spectrum in sunlight are consumed and converted.

7. Nanofluid is a reasonable way to move power and rising the output of solar systems

A stable nanocrystal line is a metal or non-metallic nanoparticle which is embedded in a base solution. Suspended nanometer particles such as silicone oxide, titanium oxide, copper oxide, or nickel metal nanoparticles, or carbon nanotubes and graphene, change fluid displacement and heat transfer properties that are extremely capable of increasing heat transfer. Nanofluid is mostly used in manufacturing because of its high coefficient of heat transfer in engines or heat exchangers to increase both performance and economy. Recently, several academic organizations and businesses have been utilizing nanofluid in solar heaters or batteries. Since the emission of light over long wavelengths on the surface of solar cells allows it to heat up, and this rise in temperature decreases performance, the cooling of solar cells is especially significant [17, 19]. At present, this cooling is achieved by going through a sea of liquid, which has no impact on temperature reduction. The usage of nanofluid moves more power from the solar cells to the outside which improves the performance which lifespan of the solar cells. At the other side, this heat may be used to preheat the water tanks and to heat the interior of the house. Nanofluid is also used in solar thermal systems. Such solar thermal devices, such as flat collectors and solar panels, solar water heaters or desalination plants, operate by collecting solar thermal energy and transmitting absorbed heat to power exchangers by another stream. The heat may be used to power water tanks or houses. To this respect, nanofluid is stronger and more efficient than regular fluids to moving heat from structures to heat exchangers. Through utilizing nanofluid, the measurements of solar thermal systems can be minimized and rendered more effective, thereby raising the initial costs of constructing and sustaining such systems.

8. The application of nanotechnology-based photocatalysts in solar cells

8.1 Nanocatalysts are a new window into improving efficiency

Photocatalysts are typically stable semiconductor oxides creating an electron-hole pair by collecting photons. Such electron holes will interfere with the molecules on the surface of the particles. Photocatalysts are used in solar panels, water purifiers, air pollution, self-cleaning lenses, decomposition of organic compounds, and so on. The strong absorption potential of photocatalysts and their susceptibility to visible and ultraviolet light have increased their spectrum of use. A number of nanophotocatalyst have been used in this respect, such as titanium dioxide, zinc oxide, cadmium sulfide, etc. The biggest issue for photocatalysts is the accumulation of small wavelengths of sunlight [21]. As a consequence, their productivity and usefulness will decline and economic costs will rise. To address the issue and consume longer wavelengths (in the spectrum of visible light wavelengths) by photocatalysts, mix them with one another or use two forms of catalysts concurrently. For example, the application of silver nanoparticles to titanium oxide is the role of the titanium oxide photocatalyst in the absorption of wavelengths.

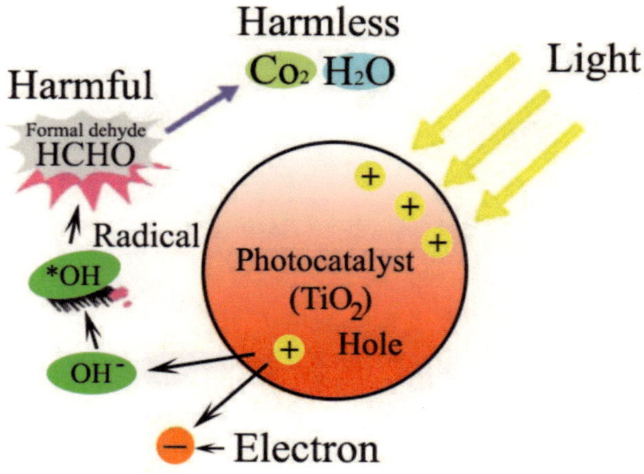

Figure 9.
Self-cleaning process in nanophotocatalysts [22].

This has greatly expanded the size from 400 to 450 nm. Considering the fact that photocatalysts absorb specific light spectra, their usage in solar cells improves the absorption of light within the cell and therefore enhances the performance of the solar cell. Often, most nanophotocatalysts have self-cleaning, anti-steam and anti-dust properties as showed in **Figure 9**, and utilizing them outside and in the body of solar cells provides an atmosphere free of air pollutants and obstacles to light in the cell and improves the absorption of sunlight and cell performance. Another function of nanophotocatalysts in solar cells, in addition to raising the spectrum of absorption and guiding it to visible light, is to enhance and increase the transition of electrons to the electrodes, thereby growing the resistance inside the cells [23]. In this situation, the recombination of the electrons with the cavities decreases and the electrical current produced rises and the energy transfer capacity improves.

9. Self-cleaning and anti-reflective nanocoatings, protect and increase efficiency

Environmental factors such as the absorption of light from the cell-solar surface, rainy weather and the barriers produced by the movement of light, such as sedimentary deposits on the surface of solar cells, are one of the factors that limit the performance of solar cells [24]. Advances in technology and the creation of nanometer layers with fascinating self-cleaning and anti-reflective properties improve the generation of solar power to address this issue. Titanium oxide nanoparticles, which can trap organic compounds, such as hydrocarbons, through blocking the ultraviolet wavelength of sunlight, may hold solar cell surfaces clear through reducing fossil fuel emissions and keeping them from being dirty (**Figure 10**).

In this way, the sunlight enters the cell surface and the reaction will be more effective and the development of electrons and cavities will begin to be more productive. In fact, utilizing nanotechnology, the hydrophilic and hydrophobic characteristics of the glass surface may be modified in such a manner that the water does not appear to damp the surface and that the sedimentary effects of the salts in the water stay on the glass surface [5]. As the amount of electron-hole output in the semiconductor cells is proportional to the strength of the sunlight, the elimination of the reflecting part of the sunlight by the protective glass of the solar cell surface and its transition to the semiconductor surface is one of the ways to increase

Figure 10.
Self-cleaning and anti-reflective nanocoatings [7].

capacity. Throughout this respect, anti-reflective nanocoatings composed of nanostructures such as nanocraft made of Polydimethylsiloxane (PDMS) or silica nanometer pores made of titanium oxide nanoparticles have been mentioned. Given that all of these nanocoatings are only in the laboratory phases of growth, increasing the performance of solar cells would render them desirable for commercialization.

10. Application of nanotechnology in power storage systems

Some of the issues with solar power generation systems are volatility and intermittent development. Power production in such systems relies on environmental factors such as atmospheric patterns, temperature, sunshine hours, so on. For this reason, continuous and consistent output in such processes is not feasible [25]. The existence of a storage unit, such as a pump, is also required in order to adjust the power supply at the moment.

Disable the intake. Ordinary batteries have a heavy weight, capacity and poor performance, so they can be expensive for the user to fix so remove. Lithium batteries are of concern in the latest wave of batteries [24]. Nanotechnology is also commonly used in this field. The most significant distinction between traditional cells and lithium batteries is the usage of organic solvents as an electrolyte solution instead of gas. In the case of lithium batteries, the lithium-ion battery creates an electrical connection between the two electrodes which, in the case of the two electrodes, transfers electrons through charging which unloading. The electrolytes used in LiPF6-based lithium batteries are primarily lithium alkyl carbonate, lithium alkoxide and other salt elements such as lithium fluoride. Some of the key issues of liquid electrolytes are the strong electrical resistance owing to the use of organic solvents. Nanomaterials are used to boost the efficiency of the electrolyte. Adding powders, particularly in the form of nanoparticles, from compounds such as aluminum oxide, silicone oxide and zirconium oxide to non-aqueous electrolytes, can increase conductivity by up to 6-fold. Extensive work has contributed to the production of solid polymer electrolytes rather than liquid first generation lithium batteries. Reducing the possibility of electrolyte contamination, growing fire tolerance and thus growing protection are properties of polymer electrolytes [26]. **Figure 11** indicates charge and discharge of a model lithium battery when connected to Photovoltaic Systems.

A great deal of concern is given to strong polymer with a polyethylene oxide foundation in the conductors of lithium ion conductors. Low price, strong chemical consistency and high protection are the hallmarks of these polymers, but the conduction of these polymers to lithium happens only at temperatures above 70° and the process of conduction in these polymers is primarily attributed to the movement of anions and the small volume of lithium transported by these polymers, which reduces the strength of the lithium-ion batteries [27]. Adding ceramic filler nanoparticles such as titanium oxide, aluminum oxide and silicone oxide to the polymer matrix polymer matrix dramatically removes this issue. Furthermore, the presence of nanoparticles inhibits the crystallization of recycled polymer chains at 70°C, which stabilizes the amorphous process at lower temperatures and improves the ion conductivity. A great deal of research has recently been conducted on crystalline oxide polymer polymers with SbLiX (where x = As, P) suggesting that such polymers contribute to lithium ions. In such polymers, the polyethylene oxide tube, in the shape of a pipe, enables the movement of lithium ions. **Figure 11** shows the formation of nanometer channels in the polymer structure which shape the lithium-ion transmission pathway (**Figure 12**).

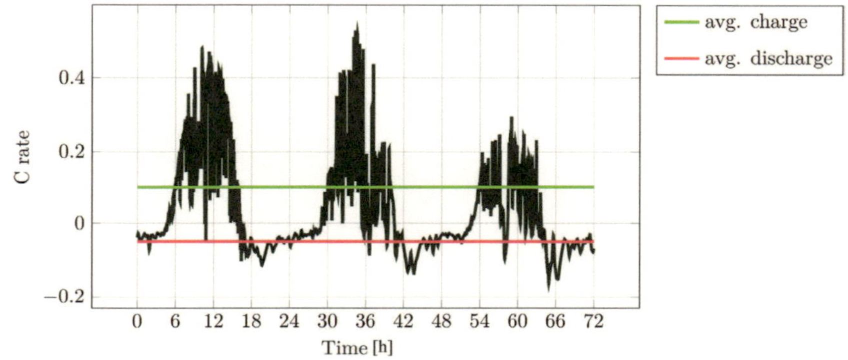

Figure 11.
One day charge and discharge of lithium batteries [26].

Figure 12.
The bank of batteries [5].

11. Conclusion

In sum, it can be seen that while the usage of nanotechnology in the construction and enhancement of solar cell efficiency is currently in the research process, it can be assumed that the transition period to the commercial arena for this field

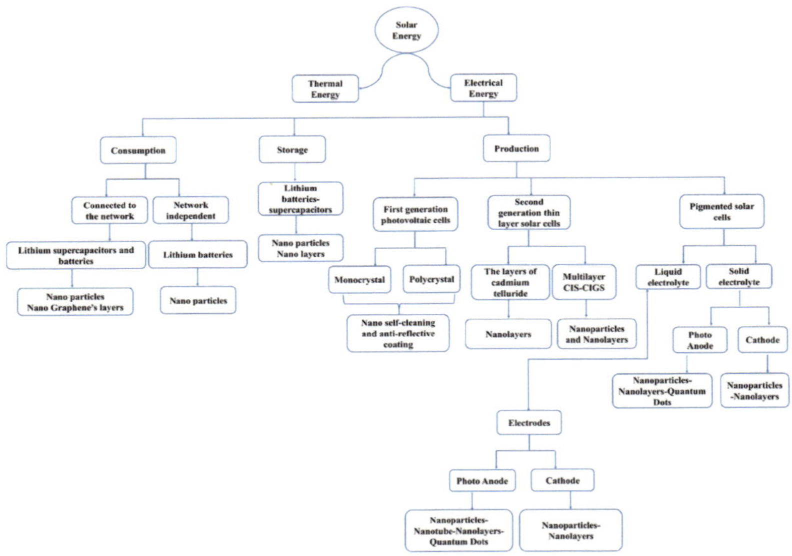

Figure 13.
The applications of nanotechnology in solar energy systems [7].

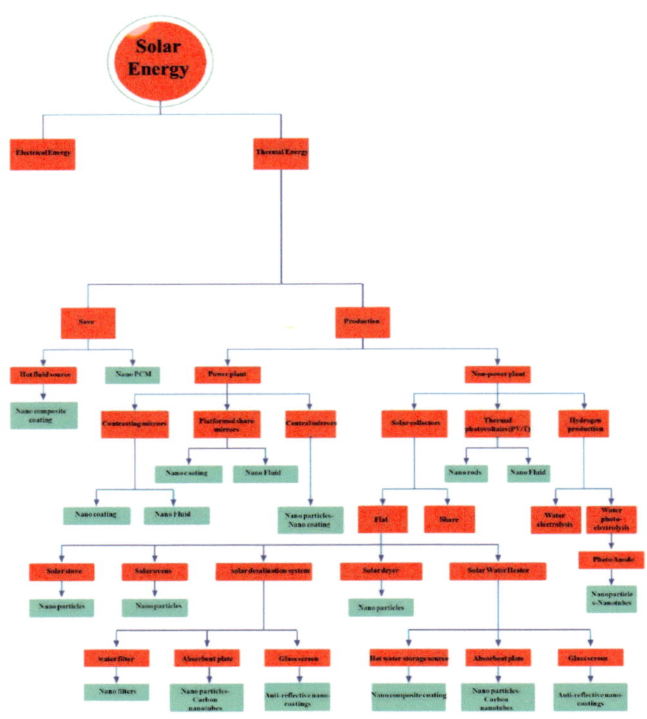

Figure 14.
Nanotechnology and solar energy tree guidance [7].

would be very near and inevitable. Seeing the tremendous promise that this sector has demonstrated in enhancing the efficiency of solar cells, the commercialization of this technology can be viewed as a major turning point in the solar cell industry. **Figures 13** and **14** show the summery of application of nanotechnology in solar technology.

Author details

Farzaneh Ghasemzadeh[1*] and Mostafa Esmaeili Shayan[2]

1 Iran University of Science and Technology, Tehran, Iran

2 Tarbiat Modares University, Tehran, Iran

*Address all correspondence to: fgh.7192@yahoo.com

References

[1] Fan Z, Wang D, Yuan Y, et al. A lightweight and conductive MXene/graphene hybrid foam for superior electromagnetic interference shielding. Chemical Engineering Journal. 2020;**381**:122696

[2] Salem MR, Elsayed MM, Abd-Elaziz AA, Elshazly KM. Performance enhancement of the photovoltaic cells using Al_2O_3/PCM mixture and/or water cooling-techniques. Renewable Energy [Internet]. August 2019;**138**:876-890. Available from: http://dx.doi.org/10.1016/j.renene.2019.02.032

[3] Panchabikesan K, Swami MV, Ramalingam V, et al. Influence of PCM thermal conductivity and HTF velocity during solidification of PCM through the free cooling concept—A parametric study. Journal of Energy Storage. 2019;**21**:48-57

[4] Otanicar T, DeJarnette D, Society of Photo-optical Instrumentation Engineers. Solar energy harvesting: How to generate thermal and electric power simultaneously. Bellingham, Washington, USA: SPIE Publications; 2016

[5] Esmaeili MS, Najafi G. Energy-economic optimization of thin layer photovoltaic on domes and cylindrical towers. International Journal of Smart Grid. 2019;**3**:84-91

[6] Murray RL, Holbert KE. Nuclear Energy: An Introduction to the Concepts, Systems, and Applications of Nuclear Processes. Amsterdam, Netherlands: Elsevier; 2019

[7] Esmaeili Shayan M, Najafi G, Gorjian S. Design Principles and Applications of Solar Power Systems (In Persian). 1st ed. Tehran: ACECR Publication-Amirkabir University of Technology Branch; 2020

[8] Bondavalli P. The graphenes cousins from dream to reality. In: Graphene and Related Nanomaterials. Amsterdam, Netherlands: Elsevier; 2018. pp. 103-136

[9] Esmaeili Shayan M, Najafi G, Ahmad BA. Power quality in flexible photovoltaic system on curved surfaces. Journal of Energy Planning and Policy Research. 2017;**3**:105-136

[10] Mahmoudi T, Wang Y, Hahn YB. Graphene and its derivatives for solar cells application. Nano Energy. 2018;**47**:51-65

[11] Kalogirou SA. Solar thermal collectors and applications. Progress in Energy and Combustion Science. 2004;**30**:231-295

[12] Lee NA, Gilligan GE, Rochford J. Solar energy conversion. Green Chemistry. 2018:881-918

[13] Singh H, Tayal PK, Verma S, et al. The third generation dye-sensitized solar cell. Renewable and Sustainable Energy Reviews. 2013;**2**:452-458. Available from: http://www.ijates.com/images/short_pdf/1459266369_381N.pdf [Accessed: 15 February 2018]

[14] Sarbu I, Sebarchievici C, Sarbu I, et al. Solar collectors. Sol Heating & Cooling System. 2017:29-97

[15] Ghasemzadeh F, Kanjouri F. Strain effect on the electronic properties of III-nitride nanosheets: Ab-initio study. Science China Technological Sciences. 2018;**61**:535-541

[16] Shakeri MG, Farzaneh Afkani M, Kanjouri F. Electronic and optical properties of Penta-BP5 monolayer. In: 7th International Congress on Nanoscience and Nanotechnology (ICNN2018); Tehran; 2018

[17] Rahbar K, Riasi A, Khatam Bolouri Sangjoeei H, et al. Heat recovery

of nano-fluid based concentrating photovoltaic thermal (CPV/T) collector with organic Rankine cycle. Energy Conversion and Management. 2019;**179**:373-396

[18] Genedy M, Kandil UF, Matteo EN, et al. A new polymer nanocomposite repair material for restoring wellbore seal integrity. International Journal of Greenhouse Gas Control. 2017;**58**:290-298

[19] Rehan MA, Ali M, Sheikh NA, et al. Experimental performance analysis of low concentration ratio solar parabolic trough collectors with nanofluids in winter conditions. Renewable Energy. 2018;**118**:742-751

[20] Tiwari GN. Handbook of Solar Energy. Singapore: Springer Singapore; 2016. DOI: 10.1007/978-981-10-0807-8

[21] Kuang A, Zhou T, Wang G, et al. Dehydrogenation of ammonia borane catalyzed by pristine and defective h-BN sheets. Applied Surface Science. 2016;**362**:562-571

[22] Milanese C, Jensen TR, Hauback BC, Pistidda C, Dornheim M, Yang H, et al. Complex hydrides for energy storage. International Journal of Hydrogen Energy [Internet]. March 2019;**44**(15):7860-7874. Available from: http://dx.doi.org/10.1016/j.ijhydene.2018.11.208

[23] Asim N, Mohammad M, Badiei M. Novel nanomaterials for solar cell devices. In: Nanomaterials for Green Energy. Amsterdam, Netherlands: Elsevier; 2018. pp. 227-277

[24] Walker A. Solar Energy: Technologies and Project Delivery for Buildings. Hoboken, New Jersey, USA: Wiley; 2013

[25] Achkari O, El Fadar A. Latest developments on TES and CSP technologies—Energy and environmental issues, applications and research trends. Applied Thermal Engineering. 2020;**167**:114806

[26] Sørensen B. Solar Energy Storage. Hoboken, New Jersey, USA: Academic Press; 2015

[27] Yan J, Luo G, Xiao B, et al. Origin of high fill factor in polymer solar cells from semiconducting polymer with moderate charge carrier mobility. Organic Electronics. 2015;**24**:125-130

Section 3

Preparation Methods of Nanoparticles: Energy Related Applications

Chapter 5

Ultrasound-Assisted Preparation Methods of Nanoparticles for Energy-Related Applications

Christos Vaitsis, Maria Mechili, Nikolaos Argirusis, Eirini Kanellou, Pavlos K. Pandis, Georgia Sourkouni, Antonis Zorpas and Christos Argirusis

Abstract

Ultrasound (US) technology is already into the research field providing a powerful tool of producing nanomaterials or being implicated in decoration procedures of catalyst supports for energy applications and material production. Toward this concept, low or/and high-frequency USs are used for the production of nanoparticles, the decoration of catalytic supported powders (carbon-based, titania, and alumina) with nanoparticles, and the production of metal-organic frameworks (MOFs). MOFs are porous, crystalline materials, which consist of metal centers and organic linkers. Those structures demonstrate high surface area, open metal sites, and large void space. All the above produced materials are used in heterogeneous catalysis, electrocatalysis, photocatalysis, and energy storage. Batteries and fuel cells are popular systems for electrochemical energy storage, and significant progress has been made in nanostructured energy materials in order to improve these storage devices. Nanomaterials have shown favorable properties, such as enhanced kinetics and better efficiency as catalysts for the oxygen reduction reaction (ORR).

Keywords: ultrasound, sonochemistry, nanomaterials, batteries, fuel cells, photocatalysis, metal-organic frameworks

1. Introduction: sonochemistry

Over the last decades, sonochemistry has been a fast developing branch of chemistry, which revolves around the ultrasound (US) effect and acoustic cavitation. USs include frequencies above the audible limit of human hearing (20 kHz). The effects of high-energy US arise from the acoustic cavitation rather than interactions of acoustic waves and matter at a molecular or atomic level. The pressure fluctuations generated by US in a liquid medium lead to the formation, growth, and implosive collapse of bubbles. More specifically, the liquid continuously expands (negative pressure) and compresses (positive pressure) until it reaches a critical diameter, which depends on the nature of the liquid and the US frequency.

The collapse of the bubble is almost an adiabatic process, and it results in a massive buildup of energy within the bubble. The microscopic bubbles can also collapse near the surface of the solid substrate and activate it, split larger particles to smaller ones or deagglomerate nanoparticles. Other than the elevated temperature and pressure, those localized hotspots can result in powerful cavitation-generated shock waves and microjets, which can cause effective stirring/mixing of the adjusted layer of liquid [1, 2].

The first region is the interior of the bubble itself, which can be visualized as a micro-/nano-reactor, dictated by extreme temperature (>5000 K) and pressure conditions (>1000 atm) along with rapid heating/cooling rates (10^{10} K/s). The second one is the interface between the bubble and the bulk solvent. The reaction efficiency of nonvolatile solutes depends on their hydrophobicity, which determines their ability to accumulate at the gas-liquid interface [3]. The third region is the vicinity of the bubble, where the bulk solution is at ambient temperature, and free radicals can form in the hot regions.

2. Batteries

As the globe encounters constant rise of energy demand to survive, the need to exploit efficiently renewable energy sources is vital. Electrochemical energy production can be a promising power source, while electrochemical devices constitute an energy storage option as well. Among these devices, batteries possess commensurate attention the last decades, due to their various applications and potential [4].

A battery cell, regardless its shape and configuration, is a device consisted of two electrodes, the anode and the cathode, an electrolyte between them having the role of the ionic conductor and a separator positioned between the electrodes that converts chemical energy to electricity or the reverse if the cell is rechargeable. Thus, the performance of the cell depends on the properties of all the components and the consistency of the system [5]. The current challenges are to advance the energy density of batteries, extend their conversion efficiency and rechargeability, and eliminate the charging time and cost while meeting the safety and environmental standards [5, 6].

In this direction, research has been emphasized toward developing nanostructured materials and implementing them in batteries as they deliver enriched performance, which is unapproachable by conventional materials. Some crucial aspects about nanomaterials are their large electrochemically active surface area, their electronic and ionic conductivity, thermal and mechanical endurance, and flexibility [7, 8]. In order to ameliorate redox reaction rates and accelerate kinetic mechanisms, a wide range of methods have been proposed including the possible incorporation of nanomaterials in each component of a cell [9].

Though the effectiveness of nanomaterials remains undisputed, only a minority of them is currently commercially utilized apparently as a result of the high cost to synthesize and manufacture them [7]. Sonochemistry may provide feasible tools to resolve many obstacles concerning the cost, safety, and environmental liability while fabricating the nanomaterials for batteries. As an example, we refer to Gu et al., who give a nice overview of graphene preparation by exfoliation in liquid media by using US as energy source and compared the results with those from other methods [10]. Graphene has excellent electrical, chemical, and mechanical properties and can, for example, boost the activity of electrodes especially of the oxygen electrodes in Zn-air batteries [11–13].

2.1 Li-batteries and Li-ion batteries

High intensity USs have been used to prepare iron-graphene hybrid electrodes for Li-batteries, whereby it was found that the ultrasonication step is of key importance for obtaining hybrid electrode material with small particle size and outstanding capacity and cyclability [14].

Olivine-structured $LiFePO_4$ is a very promising cathode material; has benefits such as nontoxicity, low cost of raw materials, and good structural stability at high temperature; is safe; and has relatively high-specific capacity (170 mA h g^{-1}) with a flat discharge-charge potential (3.45 V vs. Li^+/Li) [15, 16]. However, the low diffusion coefficients (10^{-17} to 10^{-14} cm^2 s^{-1}) of the Li-ions lead to a poor rate performance of $LiFePO_4$, and the poor electronic conductivity (10^{-9} to 10^{-8} S cm^{-1}) is hindering its practical applications and must be tackled [15–17]. The preparation of nanomaterials for cathodes in Li-ion batteries (LIBs) has been proved to be one of the effective ways to overcome the problem of the slow Li diffusion and facilitate the reaction kinetics on the cathode. It is well known that porous structures can also increase the electrode/electrolyte interface area ensuring effective electrolyte permeation in cathode materials and substantially improve the performance of $LiFePO_4$ and thus of the LIBs [17, 18]. It has been demonstrated that the adoption of ultrasonic-assisted impinging stream reaction is an effective method to produce positive electrode precursor particles for LIBs with high electrochemical performance [19].

$(NH_4)Fe_2(PO_4)_2(OH)\cdot 2H_2O$ has been prepared as positive electrode material for LIBs. The ultrasonication step was followed by a hydrothermal treatment step [20].

Apart from $LiFePO_4$, the preparation of Li-Mn-O electrode materials for secondary Li-batteries has been examined. Kim et al. fabricated mesoporous $LiMn_2O_4$ nanospheres with upgraded properties and observed their performance in 50 cycles [21]. Sonochemical treatment of the surface of a corresponding Li-Mn-O electrode by coating particles with a porous film of MgO enhances its electrochemical properties, especially in high temperatures [22].

Mn_3O_4 is used as a precursor in the preparation of $LiMn_2O_4$ and can be synthesized using US in a direct step. Co_3O_4 used as key material in energy applications can be prepared by the same procedure [23]. Co_3O_4 nanoparticles (NPs) can be obtained by sonochemical synthesis also out of ionic liquids or azo ligands, using $Co(CH_3COO)_2\cdot 2H_2O$ as starting material [24, 25]. In both reports, the diameter of the particle size of Co_3O_4 could be decreased lower than 50 nm.

A plethora of carbon morphologies has been extensively investigated as potential material appropriate for anode electrode in LIBs. Carbon spheres constitute an example applicable in LIBs [26, 27]. It has been demonstrated that the use of US provides an opportunity to prepare the mentioned material in a nontoxic accessible manner under mild conditions and competent dimensions (150–400 nm) [28].

Furthermore, recently Kumar et al. [29] outlined the progress in sonochemical synthesis of carbon dots, while Gedanken et al. [30] presented an advanced hybrid electrode of Cu foil coated by a layer of Sn@C-dots@Sn NPs ranging from 50 to 200 nm. These NPs were formed via sonication and contributed in promising cycling endurance of the cell.

A novel approach for the preparation of electrodes is the use of composite materials based on graphene. Therefore, Fe(III) oxide was sonochemically coprecipitated on graphene nanosheets in order to obtain nanocomposites for rechargeable Li batteries with stable charge-discharge kinetics for ca. 120 cycles [31]. Wu et al. also prepared magnetite NPs on reduced graphene by using a one-pot US-assisted method. These nanocomposites allow for high performance lithium ion storage devices [32].

Reduced graphene oxide (RGO) nanosheets dispersed under ultrasonic irradiation in NV ($Ni_3(VO_4)_2$) NPs prove to eliminate their agglomeration; thus, the highly conductive electrode fabricated by the composite NV/RGO can preserve 88% of its initial capacity (117.22 mA h g^{-1}) after 1000 cycle tests [33].

Polyacrylonitrile (PAN) can be used as a component of solid composite electrolyte lithium battery [34]. A sonochemical method has been used to prepare negative electrode materials containing encapsulated intermetallic NPs in PAN. The use of US leads to very small particles of $CoSn_2$, which favors the formation of amorphous Li-Co-Sn and $CoSn_2$ alloys, while the carbonaceous matrix helps to maintain the small particle size. The resulting $CoSn_2$-carbonaceous phase electrode ($CoSn_2$@C) shows improved electrochemical behavior and is stable upon cycling (ca. 450 mA h g^{-1} after 50 cycles) in comparison with reports on pure crystalline $CoSn_2$ [35].

Cu_2O-based graphene composites have been prepared and tested for use as anode materials in lithium ion batteries [36] with superior performance than Cu_2O and can be used also for photocatalytic applications, sensors, and energy storage, especially for supercapacitors [37].

Various reports investigate copper (II) oxide (CuO) and CuO/carbon composites as a possible anode material for LIBs [38–41]. Studies exploiting sonochemical methods for fabricating these materials involve the synthesis of copper complexes in an ultrasonic bath and the following calcination of compounds between 400 and 500°C. In this way, CuO particle size can be decreased until 12.1 nm. However, the specific structure and dimensions of particles differ depending on the precursor and the treatment conditions [42–44]. Hajnorouzi proposed a new method called "direct sonoelectrochemistry" incorporating ultrasonic irradiation in the process of electrolysis of a Cu tip and a physical top-down method, "US ablation" with Cu foil as the starting material to produce CuO NPs [45]. In comparison with conventional electrochemical methods, the produced amount of NPs was increased, and their dimensions were controlled, while the total time of preparation was reduced.

Nanoporous silicon structures are considered to be an attractive material in the design of LIBs as they have a large theoretical specific capacity [46–48]. Bedini et al. reported the synthesis of hydrogenated amorphous Si NPs under ultrasonic irradiation in mild conditions [49]. The product was highly porous with dimensions of particle ranging from 1.5 to 50 nm.

Two-dimensional molybdenum disulfide (MoS_2) NPs have high potential implementation not only in LIBs but also in sodium-ion [50], Li-sulfur [50], zinc-ion [50, 51], and Mg batteries [52]. Liquid-phase ultrasonic exfoliation method can be an attractive process to disperse nanosheets of MoS_2 in various solvents [53]. One more layered 2D material that can be obtained with the contribution of US is V_2O_5 nanosheets according to Li et al., who fabricated and evaluated the electrochemical performance of the respective electrode [54].

Among the oxides of manganese, manganese dioxide (MnO_2) finds application as energy storage material in alkaline batteries, rechargeable lithium batteries, and dry cells. Highly dispersed and nonagglomerated nano a-MnO_2 with a needle form of 1–2 nm diameter and up to 50 nm length have been synthesized by ultrasonication of an aqueous manganese(III)acetate solution with pH close to 7 followed by mild drying [55]. Reduction of $KMnO_4$ has been also investigated as an alternative manner to exploit ultrasonic irradiation to prepare MnO_2 [56, 57]. Okitsu et al. [56] provided useful data in order to comprehend the mechanism of basification, assisted by H_2O_2 molecules formed during sonication, while Gnana Sundara Raj et al. [57] used also polyethylene glycol so as to achieve reduction and prepare spherical MnO_2 particles with dimensions from 10 up to 20 nm. This proved to exhibit proper electrochemical endurance (after 500 cycles, 87% of the initial capacitance was preserved, while in the end of 1000 cycles, 78% of the initial specific capacitance was preserved).

Not always is it possible to obtain the result one is looking for, during the preparation of nanomaterials using US. Ganesh Kumar et al. did not obtain lithiated manganese oxide suitable for lithium batteries by treating Mn(III) salts using US and hot-hydrolysis, but the study showed the superiority of the ultrasonication regarding the adjustment of particle properties [58]. In another attempt of the same group, $LiNi_{0.5}Mn_{1.5}O_4$ was synthesized sonochemically as cathode with high redox potential for LIBs with better cyclability [59].

2.2 Zn-air and Zn alkali batteries

Zinc-air batteries (ZABs) exhibit a high energy density being at the same time a low-cost product. Therefore, a tremendous interest is present in meeting the demands for flexible and portable electronics. A novel porous-structured poly vinyl alcohol (PVA)-based nanocomposite gel polymer electrolyte (GPE) with silica (SiO_2) was synthesized and used as electrolyte in a flexible ZAB. The fabricated porous material exhibited a high ionic conductivity (57.3 mS cm^{-1}), excellent water retention capability, and improved thermal and mechanical properties under ambient condition, and the ZAB showed an excellent cyclability, discharge performance, and power density [60].

It has been reported that cerium metallic particles deposited on Zn anode for alkali batteries lead to an improved electrochemical performance, whereby US power and sonication time influence strongly the battery efficiency, increase the corrosion resistance of the anode, and suppress the Zn-dendrite formation [61].

Regarding the oxygen reduction reaction (ORR), highly efficient cathodes for ZABs have been prepared based by decorating Fe_2P on 3D N,P-codoped porous carbon. The later has been prepared using pore-forming agents [62]. Further, bimetallic oxides like perovskites can be immobilized on different substrates and used as air electrodes in ZABs. As an example, one can use nafion, which has anti-fouling properties and is very interesting in electrochemical application owing to its interesting electronic and catalytic properties. As an example, we refer to Chen et al., who have immobilized $SrWO_3$ on nafion by using US [63].

Nickel-iron layered double hydroxide (NiFe LDH) constitutes one more competitive catalyst with potential use in air cathodes, due to its layered structure. Sonication-assisted liquid exfoliation has been proposed to be competent to deliver highly functional NiFe LDH/CB nanosheets considering their oxygen evolution (OER) catalytic properties and stability [64].

2.3 Other battery systems

Xie et al. prepared active cathodes for $Ag_2V_4O_{11}$/Li battery systems through a reaction between V_2O_5 gel and Ag_2O powder, which has been accelerated by using US [65]. The prepared cathodes ($Ag_2V_4O_{11}$ and $Ag_{1.4}V_3O_8$) exhibited superior electrochemical properties as compared to the ones prepared by this solid-state method.

In Li-S batteries, the sulfur host plays an important role. With respect to this, hierarchically ordered micro/mesoporous carbon (HPC) has been prepared by US-assisted spray pyrolysis obtaining HPC-S cathodes, which exhibited an excellent cycle retention of 77% in tests with 500 cycles at 2.4°C [66, 67].

3. Fuel cells

An electrochemical mechanism for the direct combustion (chemical oxidation) of fuels, which bypasses the intermediate stage of heat generation (so-called "cold combustion"), allows for the direct conversion of chemical energy of a fuel to electrical energy without the losses due to the Carnot process. This direct energy conversion is possible in devices called fuel cells, and the underlying electrochemical mechanism has analogies in living beings [68].

There are several types of fuel cells, mostly categorized based on the used electrolyte. A second classification is based on the temperature needed by the electrolyte to achieve sufficient ionic conductivity and one speaks about low-, intermediate-, and high-temperature fuel cells.

3.1 Low-temperature fuel cells

The most important reaction in fuel cells is the provision with enough oxidant in order to keep the reaction running and thus the electricity production at the maximum level. This reaction is the reduction of the oxygen molecule (oxygen reduction reaction—ORR) as this is the source of the ions either to be transported through the solid electrolyte in solid oxide fuel cells (SOFCs) or to react with the protons arriving through the electrolyte in polymer electrolyte membrane (PEM) fuel cells.

In PEMs, the operating temperature is low, and therefore, the kinetics of the ORR is not high enough without the use of catalysts [69].

We prepared using the sonoelectrochemical method a Pt and carbon black-based nanocomposite as electrocatalyst for PEM fuel cells. We used pulsed electrodeposition in combination with pulsed ultrasonication to obtain Pt NPs on carbon black substrates, and we have shown the beneficial role of polyvinylpyrrolidone (PVP) against the agglomeration of the produced NPs [70].

Despite the reports of many research groups that Pt-based materials are considered to be the best electrocatalyst for ORR in fuel cells, there is no doubt that their reserves in nature are very limited making them expensive. Further, their durability is not as high as needed to use them in commercial applications. Therefore, reduction of the Pt consumption and most importantly their replacement with nonprecious metal catalysts in the ORR are considered essential. Therefore, the development of non-Pt or metal-free ORR electrocatalysts is extremely important and urgent [71].

Pd-based catalysts are one of the most attractive choices for the replacement of Pt catalysts as their cost is significantly lower, and at the same time, they possess a high catalytic activity for ORR not only because of the Pd itself but also due to synergistic effects between the Pd and the other components and supports [72]. Until now, different Pd-based composite materials have been proposed and investigated as catalysts for ORR [73, 74].

A very useful overview on sonochemically prepared multicomponent electrocatalytic materials for low-temperature fuel cells is given by Lee and Kwon [75]. Most of the efforts are focusing on the partial replacement of Pt by low-cost metals, for example, Ni [76], or the Pt replacement by Pd and respective nanoalloys with low-cost metals (Mn and Fe as core-shell with Pd) as well [77]. Carbon-supported Sn NPs for electrochemical applications and especially for improving the kinetics of the ORR have been sonochemically synthesized and showed high reduction overpotential for the ORR mainly due to the high surface area of the resulting carbon-supported Sn electrode [78].

Further improvement in the catalytic activity of the developed catalysts is expected through core-shell architecture materials and also through the use of active supports with high porosity leading to high active centers on the catalyst surface. Unique Pd@Pt/C core-shell NPs as methanol-tolerant catalysts have been prepared by Zheng et al. in a sonochemical multistep approach [79]. The high performance of the Pd_3Pt/C catalyst is ascribed to the unique combination of preferable growth of the Pd (1 1 1) plane, small particle size (~4 nm), unique core/shell structure, and the electronic effects between Pd and Pt.

Alternative electrocatalytic materials for the ORR have been also prepared starting from biomass and turned out to be promising alternatives to noble metal catalytic materials. The resulting catalysts exhibit an excellent catalytic activity as compared to commercial catalysts with reduced methanol crossover [80].

Also, oxides of transition metals have been sonochemically prepared as electrocatalysts for the ORR. Highly active porous MnO_2 with superior electrocatalytic activity as compared to commercial Pt/C catalyst has been sonochemically prepared and tested by Zuo et al. as a promising catalyst for direct methanol fuel cells [81].

US has been used also for the preparation of electrolyte membranes for PEMs. Nanocomposite membranes based on sulfonated polybenzimidazole (PBI) with cellulose and silica precursors have been made with improved mechanical properties and decreased methanol permeability [82].

Zuo et al. prepared a composite cathode material for alkaline fuel cells based on MoS_2 decorated with Pd using a simple sonochemical route [83]. They found that the new electrocatalyst has better performance than commercial Pt/C catalysts.

3.2 High-temperature fuel cells

Fuel cells working at temperatures higher than 500°C are referred to as intermediate (<700°C) and high temperature (700–850°C) fuel cells. These are proton conducting ceramic fuel cells (PCFCs), molten carbonate fuel cells (MCFCs), and solid oxide fuel cells (SOFCs). In all these fuel cells, the ORR is important, but in general, the use of catalysts is not necessary on the cathode side because of the high service temperatures.

In SOFCs, the ORR is as important as mentioned in the PEM section, but here the temperature is high enough to accelerate the ORR without the use of specific catalysts [84]. SOFC cathodes must be efficient mixed ionic-electronic conductors (MIECs) as they need to transport both electrons and ions especially in intermediate SOFCs (IT-SOFCs) [85].

Once the cathode reaction is providing a sufficient amount of oxygen ions in SOFCs, the anode material is of outmost importance as it has to catalyze the oxidation reaction.

One of the critical components for such a device is hydrogen, which is the fuel to be oxidized. Hydrogen can be produced by not only the classical methods such as reforming of hydrocarbons, gasification of coal or heavy oil fractions, and electrolysis using renewable or nuclear energy sources but also sonochemically and sonoelectrochemically, as reported in a recent review [86]. Other groups have developed electrocatalysts for hydrogen evolution using US. High intensity ultrasonic irradiation of AlNi alloy has led to an electrocatalyst for water splitting with high surface area and changes in its composition, which can be controlled by the selection of the right fluid during sonication [87]. Nitrogen doped reduced graphene oxide supported on N-titania as efficient catalysts for the production of hydrogen through water splitting has been prepared in a combined sonochemical/hydrothermal step [88].

As oil and natural gas supply is well established, feeding SOFCs directly with natural gas would be an ideal solution [89]. For natural gas fed SOFCs, the catalytic activity of the anode materials is critical as it needs not only to accelerate the oxidation reaction but also to prevent poisoning of the active centers by coking and sulfur and to be stable against other components that may be contained in the natural gas.

In our group, several efforts have been made in order to improve the preparation methods of SOFCs [90–94] and SOFC materials [95, 96] with a focus on the anode compartment [97–99]. Emphasis was given on the implementation of US in order to reduce the preparation time or to follow a facile and/or alternative path for materials with improved properties. In most cases, we prepared nanomaterials because they offer special properties to the fuel cells as they are catalytically active to a wide range of chemical reactions. One can prepare novel SOFC anodes by decorating state-of-the-art anode powder with nanometric metals and metal oxides [100]. We decorated anode materials based on GDC/Nickel and YSZ/Nickel cermets with molybdenum and tungsten oxide NPs and improved the catalytic activity and stability of the resulting composite anodes against coking and sulfur poisoning [101].

Not only electrodes for SOFCs have been made using ultrasonication but also electrolytes. Okkay et al. prepared samaria doped ceria (SDC—$Ce_{0.8}Sm_{0.2}O_{1.9}$) using an US-assisted coprecipitation method [102]. It has been found that the lattice parameter of the produced nanomaterial increased with increasing ultrasonic acoustic power and is linearly related to the ionic conductivity of the resulting electrolyte after sintering at 1200°C. Pinjari and Pandit prepared sonochemically at room temperature ceria nanopowders with particle size less than 30 nm with clear benefits regarding energy efficiency and reaction time as compared to the conventional preparation method [103]. Sonochemistry has been also used to prepare Ce(III) nano-sized precursors for nanoceria [104].

3.3 Other electrocatalytic applications

USs can be used not only in batteries and fuel cells but also in many other electrochemical and sonoelectrochemical applications. Ultrasonication has been used for the preparation of electrocatalysts for the direct electrooxidation of ethanol. A facile US-assisted method was proposed to fabricate the Pd-Pt alloy/multiwalled carbon nanotube (Pd-Pt/CNTs) nanocomposites for the ethanol and methanol electrooxidation reaction in alkaline media [105].

In another attempt, a catalyst made of graphene supported Ag decorated Pd NPs with exceptional activity and uniformity. In this respect, it has been shown that graphene is very important as substrate as it minimizes the coalescence of the NPs, which would decrease both the surface area and the electrocatalytic activity [106]. A multifunctional nanostructured electrocatalyst has been prepared by replacing carbon copper nanowires by Pd resulting to Pd@CuNWs and supported them by multiwalled carbon nanotubes (MWCNTs) using chitosan (CH) as a binder. Electrochemical catalytic activity and durability evaluation results proved the superiority of the resulting Pd@CuNWs/MWCNTs/CH regarding electrocatalytic activity and long-term stability compared to Pd/MWCNTs and commercial Pd/C electrocatalysts for ethanol electrooxidation [107].

An overview on fundamental studies of sonochemical and sonoelectrochemical nanomaterial preparation is given in recent publications of our group on fuel cells [108, 109] and others on nanomaterials [110].

Silica gels have been considered as appropriate matrices for the preparation of complex center doped materials for a variety of applications such as controlled-release carrier implantable materials for low weight drugs in biological systems and as substitute materials for membrane processes in fuel cells [111, 112].

Ultrasonication can be used for the atomization of methanol in order to have a smooth and continuous feed in direct methanol fuel cells, leading to a high and stable open circuit voltage (OCV) [113] or to enable improvement of direct methanol fuel cells using sonication in parallel with a novel cell design with integrated ultrasonic transducer [114].

4. Photocatalysis

Photocatalysis is a phenomenon based on redox reactions, which take place at the surface of a semiconductor material under UV or visible light irradiation. The photocatalytic activity of the catalyst depends on its ability to create electron-hole pairs, which are then taking part in a redox reaction to generate hydroxyl and superoxide radicals, which are able to undergo secondary reactions [115].

The improvement of the photocatalytic efficiency is a strategy, which was developed to push the absorption onset of TiO_2 toward longer wavelengths (anatase band gap, 3.2 eV) by doping TiO_2 with anions and/or cations and metal ions [116–118]. Zinc oxide can be either a n-type or a p-type semiconductor with a wide band gap ($E_g \approx 3.3$ eV at 300 K), while its composites are very interesting materials because of possible synergistic effects on photoelectrochemical properties and photocatalytic activity [119, 120]. The primary ways to improve the photocatalytic effect can incorporate a sonocatalytic technique [121], doping [122], or stratified films [123, 124].

4.1 Environmental and energy applications

The photocatalytic degradation of organic pollutants such as dyes, pesticides, and pharmaceutical waste is a crucial application for the safety of the ecological system, mainly due to their toxicity and degradation complexity. The main application areas in catalysis are photocatalytic electrolysis of water, environmental protection, and solar cells. The pollutants in wastewater can be roughly divided into organic and inorganic pollutants, where organic compounds can be degraded by TiO_2 photocatalytic technology [125].

Converting CO_2 waste into valuable carbon fuels is undoubtedly one of the most viable and economical alternatives to reduce the CO_2 emissions and resolve the energy crisis. UV irradiation and visible light have been used as sources of excitement for semiconductor catalysts to produce energy-bearing products such as methane, methanol, carbon monoxide, formic acid, and formaldehyde. Photocatalytic reduction of CO_2 can not only reduce the carbon dioxide emissions but also solve the energy crisis [126–128]. Some of the catalysts that can be used for the photocatalytic reduction of CO_2 include WO_3 [120], $ZnIn_2S_4$ [129], CdS [130], Cu_2O [117], $CuInS_2$ [131, 132], and $BiVO_4$ [133].

4.2 Sonochemical synthesis of nanocatalysts

Stucchi et al. used sonochemistry to both form NPs from the precursor and achieve a good distribution on the TiO_2 decoration surface [117]. In fact, US energy accelerates the diffusion of the dissolved substance into the reaction system and also affects the selective adsorption of the surfactant on copper, causing elongation or compression in certain directions, thus affecting the morphology of the particles. The utilization of Cu, CuO, and Cu_2O NPs on TiO_2 surface can greatly enhance the photodegradation of acetone and acetaldehyde [118].

CdS/TiO_2 can be prepared at a relatively low temperature (70°C) with small particle sizes (11 nm) using US in a short time (1.5 h) [130]. On the other hand, the use of conventional methods requires at least 20–24 h and elevated temperatures (200–400°C). The properties of complex core-shell materials are combinations of the properties of both materials in the core and the shell. Those materials can be used in photovoltaic cells, optical sensor photocatalysts, and catalysts. In addition, CdS/TiO_2 NPs can selectively bind heavy metal ions, such as Cr (VI), on their surface [134].

The irradiation of $W(CO)_6$ in diphenylmethane in the presence of an Ar-O_2 mixture for 3 h can lead to tungsten oxide NPs consisting of both orthomolecular and monoclinic WO_2, partial oxidation of which produces tricyclic WO_3 [135]. WO_3 acts as a catalyst in reducing CO_2 in fuels (CH_4 and CH_3OH) with significant catalytic efficiency [136]. The introduction of CdS on WO_3 can enhance carbon dioxide adsorption and increase CH_4 selectivity, while the existence of two different regions can minimize undesirable back reactions of the photocatalytic products [137]. Those material CdS/WO_3 can also use for the photocatalytic degradation of organic dye rhodamine B [138].

Xin et al. synthesized $ZnIn_2S_4$ nanosheets with hexagonal and cubic structures. The samples were prepared and used to form methyl formate by photochemically reducing CO_2 to methanol. The efficiency of the hexagonal form was better than the cubic one. In addition, both hexagonal and cubic nanosheets exhibited much higher activity than $ZnIn_2S_4$ microspheres prepared by the hydrothermal method [129]. The $ZnIn_2S_4$-In_2O_3 structure is effectively used as a photocatalyst in CO_2 reduction, by offering a large surface area for CO_2 adsorption, while it exhibits abundant active sites for surface catalysis, leading to significant CO production rate and high stability [139].

Copper can also be used to prepare a $CuInS_2$ NP structure. The study of various parameters, such as the different crystallographic structures of sulfur, the concentration of precursors, the reaction time, and the power of ultrasonic radiation on the morphology and particle size, showed that the crystallinity of sulfur plays an important role in the morphology of $CuInS_2$ [131, 132]. Reducing CO_2 to solar fuel can be essential for both decreasing CO_2 emissions and increasing energy production. This photoelectrochemical reduction of CO_2 to methanol is carried out by using p-$CuInS_2$ as a photocathode [140].

$BiVO_4$ NPs can be synthesized sonochemically at room temperature at different pH values (3, 5, and 10) of the original precursor without further heat treatment. The morphologies of the final samples are different depending on the pH value of the original precursor. The $BiVO_4$ sample, which was prepared at a higher pH value, has an advantage in photocatalytic performance. The excellent photocatalytic efficiency can be attributed to the superior crystallinity and the large active surface of the $BiVO_4$ structure [133], while its photocatalytic activity was studied during the degradation of organic dyes [134].

5. Metal-organic frameworks

5.1 Properties and applications

Metal-organic frameworks (MOFs) are a new class of porous crystalline hybrid materials that have achieved a tremendous growth over the last decades, with attention not only in chemistry but also in general science and technology. They consist of inorganic metal-based centers (ions or clusters) and organic ligands, assembled through strong coordination bonds in order to create an open crystalline framework with permanent porosity.

These ordered crystalline structures possess physicochemical properties, such as high surface area, open metal sites, and large void space. The easy tuning of the shape, size, and chemical nature of pores has led to unique chemical versatility and various morphologies, such as micro- or nano-spheres, -cubes, -sheets, and -rods [141–143]. Furthermore, accessing the molecular adsorption sites has opened the way to host-guest interactions and the ability to capture materials in both chemisorption and physisorption states [144]. Due to their controllable composition and targeted preparation, MOFs can be manipulated, so they can be used in specific applications.

MOFs have been very promising in a wide spectrum of applications, ranging from the well-known gas storage/adsorption [145] and separation [146], catalysis [147], sensing [148], and dye/toxic material removal [149] to recently rising fields, such as luminescence [150], membranes [151], and drug delivery [152]. In terms of energy-related applications, they have been used for solar energy conversion [153], supercapacitors [154], batteries [155], and fuel cells [156].

5.2 Synthesis methods

MOFs have been traditionally synthesized either at room temperature [157] or via a hydrothermal/solvothermal approach by using electrical heating, an oil bath, or an autoclave at high temperatures, for a prolonged time of hours or even days. Recently, new methods have arisen to avoid these conditions. Similar to conventional, as a heating-based technique, microwaves (MWs) have lately been used widely for the synthesis of MOFs [158, 159], by offering phase selectivity, fast crystallization, and control over the crystal morphology. Electrochemical [160] and mechanochemical syntheses [161] are alternative methods, which appeared in 2005 and 2006, respectively. Other than the above popular methods, slow diffusion [162], reverse micelle [163], and combinations like sonoelectrochemical [164] have been tested as well.

5.3 MOF synthesis via US

This section is focused on the effect that US synthesis conditions have on the final product along with the use of sonochemically prepared MOFs in environmental applications with regard to harmful substance removal. $(Zn_3BTC_2)\cdot 12H_2O$ (BTC = 1,3,5-benzenetricarboxylate) was the first MOF that was successfully prepared by applying a sonochemical method in 2008 [165], followed by MOF-5 [166] and ZnBDC (BDC = 1,4-benzenedicarboxylate) [167] later that year.

In order to reveal the determining factors of the reaction rates, a kinetic study was performed in 2010, by comparing conventional, microwave, and US syntheses [168]. Fe-MIL-53 was chosen for the comparison due to its mild synthesis conditions. The average reaction time to obtain the product is 1.5–3 days at 70–80°C for the conventional synthesis, 1.5–2.5 h at 60–70°C for microwaves, and 0.5–1 h at 50–70°C for the US, which is a result of increased preexponential factors in the sonochemical method.

Synthesis conditions can heavily affect the quality of crystals, the particle size, the surface area, and the morphology; thus, various studies have been published in an effort to optimize the reaction time, US power, solvent ratios, reagent concentration, and modulators/additives and achieve the best results in targeted applications [169, 170].

5.4 US advantages on MOFs

The contamination of the environment is becoming an aggravating problem [171, 172]. As a result of the accelerated expansion of chemical, pharmaceutical, and agricultural industries, many hazardous compounds, such as dyes, antibiotics, and pesticides, reach the aquatic environment. Therefore, many scientists have dedicated their work for the production of materials that can remove harmful substances.

Abbasi et al. compared the sonochemical synthesis of HKUST-1 with a mechanochemical one [173]. The majority of the mechanochemically prepared particles had a size of at least 60 nm, while the US led to particles mainly around 35–65 nm. Their adsorption efficiency was tested via the removal of two dyes. After 24 h, 19.52% of methylene blue and 10.86% of crystal violet had been adsorbed by the mechanochemical MOFs, while the US one had managed to remove 31.91 and 27.43%, respectively. The increased adsorption of methylene blue could be explained due to its smaller size, so it can be captured more easily than crystal violet.

In order to improve the methylene blue adsorption performance of TMU-23, a composite was prepared with graphene oxide (GO) in a US bath within 60 min at room temperature [174]. About 30 mg of GO-TMU-23 (10% GO) was added in 20 mL of 10 ppm aqueous solution, as well as 27 mg TMU-23 and 3 mg GO were also examined separately for comparison purposes. After only 2 min, TMU-23, GO, and GO-TMU-23 have removed 50, 47, and 89% of MB, respectively, while after 15 min, they have removed 78, 90, and 97%, respectively. Another GO composite is GO-Ni-BTC, which was prepared via an US-assisted ball milling technique [175]. Water was used as the only solvent as the coupling effect of mechanical force and ultrasonic waves can promote the reaction without an organic solvent. The Ni-BTC and GO-Ni-BTC were compared by studying the thermodynamics, along with adsorption kinetics by using the congo red dye, resulting in capacities of 2046 and 2489 mg/g, respectively.

$[Zn(ATA)(BPD)]_\infty$ nanoplates (ATA = 2-aminoterephthalic acid, BPD = 1,4-bis(4-pyridyl)-2,3-diaza-1,3-butadiene) were prepared at room temperature in a US bath within the optimized duration of 120 min [176]. The addition of triethylamine accelerated the crystal growth, and crystals could get separated after 60 min, but 120 min was still needed to receive fully uniform nanoplates. The modulator implementation led to an increase in 2,4-dichlorophenol adsorption from 68 to 91%. Pyridine had also been tested as a modulator agent during the synthesis of $[Zn(TDC)(4\text{-}BPMH)]_n{\cdot}n(H_2O)$ (TDC = 2,5-thiophene dicarboxylic acid, 4-BPMH=N,N-bis-pyridin-4-yl-methylene-hydrazine) [177]. Its addition led to better morphology of uniform sheet-shaped nanoplates with a thickness of 20–35 nm instead of NPs without order, while it also increased the removal efficiency of dichlorophenol and amoxicillin from 92.5 to 95% and from 87 to 94.5%, respectively, after 3 h.

The adsorption of rifampicin (antibiotic drug) [178] and imatinib (anti-cancer drug) [179] was tested as well with HKUST-1. The MOF was prepared by both a sonochemical method within 60 min and a conventional one after 24 h at 80°C. The US synthesized particles had an average size of 80 nm, while the majority of the conventional ones had a size over 150 nm. Regarding the uptake properties, the former could adsorb 26.6% of rifampicin after 3 h and 98% after 48 h, while the latter could reach 19% and 59.6%, respectively. Similarly, although a greater time period was needed, after 144 h, 96.7% of imatinib was adsorbed by the US HKUST-1 and 81.1% by the conventional one.

6. Conclusions

Among the several types of fuel cells, the use of nanoparticles has been implemented for improving the ORR in cathodes and the stability and poisoning of the anodic catalysts enhancing mainly the mass transfer phenomena. US-assisted preparation methods have been enabling more facile and cost-effective preparation methods of producing mono- and bimetallic nanoparticles in the absence and presence of various surfactants than conventional methods of preparations. The produced nanoparticles were decorated onto cathode surfaces used in low- and high-temperature fuel cells acting synergistically with the surface toward an enhanced catalytic activity. The use of US and/or sono(electro)chemical methods also produces a controllable variety in the shape of the exposed planar sites of the moieties onto electrode surfaces and was proven to be a powerful tool for reducing metal precursors to mono- or/and bimetallic nanoparticles. Especially, in the case of carbon-supported nanocatalysts, the use of US, in general, has been proposed toward a high-value product production to meet the needs of energy applications.

US-assisted preparation methods are also a very promising tool in developing highly efficient materials for batteries. Significant efforts have been focused on creating functional nanomaterials in a variety of morphologies while decreasing the time and cost of preparation for meeting the commercial requirements. Substantial research has been reported in the field of Li and Li-ion batteries, where certain materials were directly tested in custom-made batteries with promising performance. Concerning other battery systems such as zinc-based and metal-air batteries, research showed promising results, although various nanomaterial candidates for electrodes, electrolytes, or catalysts have been prepared but not thoroughly tested. A more focused research trend in implementing the synthesized US-assisted nanomaterials in battery applications is currently toward a complete range definition of characteristics and stability in order to overcome possible failures and limitations.

In terms of environmental studies, semiconductor nanoparticles and composites prepared via US technology have shown promising results toward photocatalysis. Carbon dioxide reduction has achieved great process on both the reaction mechanisms and the pathways; thus, such materials can display better product selectivity.

Finally, USs have opened the way to fast and facile synthesis of metal-organic frameworks by further offering smaller particle size and enhanced morphologies. Compared to conventional methods, the accelerated nucleation and crystallization times have made sonochemistry very attractive over the last few years, although proper handling is necessary in regard to synthesis conditions for the acquisition and optimization of desired properties.

Author details

Christos Vaitsis[1], Maria Mechili[1], Nikolaos Argirusis[2], Eirini Kanellou[1,4], Pavlos K. Pandis[1], Georgia Sourkouni[3], Antonis Zorpas[4] and Christos Argirusis[1,3]*

1 School of Chemical Engineering, National Technical University of Athens, Zografou, Greece

2 mat4nrg – Gesellschaft für Materialien und Energieanwendungen mbH, Germany

3 TU Clausthal, Clausthaler Zentrum für Materialtechnologie, Germany

4 Faculty of Pure and Applied Sciences, Open University of Cyprus, Nicosia, Cyprus

*Address all correspondence to: amca@chemeng.ntua.gr

References

[1] Mason TJ, Lorimer JP. Applied Sonochemistry: The Uses of Power Ultrasound in Chemistry and Processing. Weinheim: Wiley; 2002. p. 303. DOI: 10.1002/352760054X

[2] Suslick KS. Sonochemistry. Science. 1990;**247**(4949):1439-1445. DOI: 10.1126/science.247.4949.1439

[3] Henglein A. Sonochemistry: Historical developments and modern aspects. Ultrasonics. 1987;**25**(1):6-16. DOI: 10.1016/0041-624x(87)90003-5

[4] Winter M, Brodd R. What are batteries, fuel cells, and supercapacitors? Chemical Reviews. 2004;**104**:4245-4269. DOI: 10.1021/cr020730k

[5] Linden D, Reddy T. Handbook of Batteries. New York: McGraw-Hill Education; 2001

[6] Ryu J, Song WJ, Lee S, Choi S, Park S. A game changer: Functional nano/micromaterials for smart rechargeable batteries. Advanced Functional Materials. 2019;**30**(2):1902499. DOI: 10.1002/adfm.201902499

[7] Pomerantseva E, Bonaccorso F, Feng X, Cui Y, Gogotsi Y. Energy storage: The future enabled by nanomaterials. Science. 2019;**366**(6468). DOI: 10.1126/science.aan8285

[8] Ibrahim ID, Sadiku ER, Jamiru T, Hamam Y, Alayli Y, Eze AA. Prospects of nanostructured composite materials for energy harvesting and storage. Journal of King Saud University - Science. 2020;**32**(1):758-764. DOI: 10.1016/j.jksus.2019.01.006

[9] Kumar P, Kim K-H, Bansal V, Kumar P. Nanostructured materials: A progressive assessment and future direction for energy device applications. Coordination Chemistry Reviews. 2017;**353**:113-141. DOI: 10.1016/j.ccr.2017.10.005

[10] Gu X, Zhao Y, Sun K, Vieira CLZ, Jia Z, Cui C, et al. Method of ultrasound-assisted liquid-phase exfoliation to prepare graphene. Ultrasonics Sonochemistry. 2019;**58**:104630. DOI: 10.1016/j.ultsonch.2019.104630

[11] Bolotin KI, Sikes KJ, Jiang Z, Klima M, Fudenberg G, Hone J, et al. Ultrahigh electron mobility in suspended graphene. Solid State Communications. 2008;**146**(9-10):351-355. DOI: 10.1016/j.ssc.2008.02.024

[12] Balandin A, Bao W, Calizo I, Teweldebrhan D, Miao F, Lau J. Superior thermal conductivity of single-layer graphene. Nano Letters. 2008;**8**:902-907. DOI: 10.1021/nl0731872

[13] Lee C, Wei X, Kysar JW, Hone J. Measurement of the elastic properties and intrinsic strength of monolayer graphene. Science. 2008;**321**(5887):385-388. DOI: 10.1126/science.1157996

[14] Gonzalez JR, Menendez R, Alcantara R, Nacimiento F, Tirado JL, Zhecheva E, et al. High-intensity ultrasonication as a way to prepare graphene/amorphous iron oxyhydroxide hybrid electrode with high capacity in lithium battery. Ultrasonics Sonochemistry. 2015;**24**:238-246. DOI: 10.1016/j.ultsonch.2014.12.001

[15] Padhi AK, Nanjundaswamy KS, Masquelier C, Okada S, Goodenough JB. Effect of structure on the Fe_3/Fe_2 redox couple. Journal of the Electrochemical Society. 1997;**144**(5):1609-1613. DOI: 10.1149/1.1837649

[16] Ding Y, Jiang Y, Xu F, Yin J, Ren H, Zhuo Q, et al. Preparation of nano-structured $LiFePO_4$/graphene composites by co-precipitation method.

Electrochemistry Communications. 2010;**12**(1):10-13. DOI: 10.1016/j.elecom.2009.10.023

[17] Vu A, Qian Y, Stein A. Porous electrode materials for lithium-ion batteries - How to prepare them and what makes them special. Advanced Energy Materials. 2012;**2**(9):1056-1085. DOI: 10.1002/aenm.201200320

[18] Wang KX, Li XH, Chen JS. Surface and interface engineering of electrode materials for lithium-ion batteries. Advanced Materials. 2015;**27**(3): 527-545. DOI: 10.1002/adma.201402962

[19] Dong B, Huang X, Yang X, Li G, Xia L, Chen G. Rapid preparation of high electrochemical performance $LiFePO_4/C$ composite cathode material with an ultrasonic-intensified micro-impinging jetting reactor. Ultrasonics Sonochemistry. 2017;**39**:816-826. DOI: 10.1016/j.ultsonch.2017.06.010

[20] Dong B, Li G, Yang X, Chen L, Chen GZ. Controllable synthesis of $(NH_4)Fe_2(PO_4)_2(OH).2H_2O$ using two-step route: Ultrasonic-intensified impinging stream pre-treatment followed by hydrothermal treatment. Ultrasonics Sonochemistry. 2018;**42**:452-463. DOI: 10.1016/j.ultsonch.2017.12.008

[21] Kim JM, Lee G, Kim BH, Huh YS, Lee GW, Kim HJ. Ultrasound-assisted synthesis of Li-rich mesoporous $LiMn_2O_4$ nanospheres for enhancing the electrochemical performance in Li-ion secondary batteries. Ultrasonics Sonochemistry. 2012;**19**(3):627-631. DOI: 10.1016/j.ultsonch.2011.10.002

[22] Gnanaraj JS, Pol VG, Gedanken A, Aurbach D. Improving the high-temperature performance of $LiMn_2O_4$ spinel electrodes by coating the active mass with MgO via a sonochemical method. Electrochemistry Communications. 2003;**5**(11):940-945. DOI: 10.1016/j.elecom.2003.08.012

[23] Askarinejad A, Morsali A. Direct ultrasonic-assisted synthesis of sphere-like nanocrystals of spinel Co_3O_4 and Mn_3O_4. Ultrasonics Sonochemistry. 2009;**16**(1):124-131. DOI: 10.1016/j.ultsonch.2008.05.015

[24] Al-Qirby LM, Radiman S, Siong CW, Ali AM. Sonochemical synthesis and characterization of Co_3O_4 nanocrystals in the presence of the ionic liquid [EMIM][BF_4]. Ultrasonics Sonochemistry. 2017;**38**:640-651. DOI: 10.1016/j.ultsonch.2016.08.016

[25] Cheng XX, Hojaghani S, Hu ML, Hosaini Sadr M, Morsali A. Sonochemical synthesis and characterization of new nanostructures cobalt(II) metal-organic complexes derived from the azo-coupling reaction of 4-amino benzoic acid with anthranilic acid, salicylaldehyde and 2-naphtol. Ultrasonics Sonochemistry. 2017;**37**:614-622. DOI: 10.1016/j.ultsonch.2017.02.007

[26] Pol V, Thackeray M. Spherical carbon particles and carbon nanotubes prepared by autogenic reactions: Evaluation as anodes in lithium electrochemical cells. Energy & Environmental Science. 2011;**4**:1904-1912. DOI: 10.1039/c0ee00256a

[27] Roberts AD, Li X, Zhang H. Porous carbon spheres and monoliths: Morphology control, pore size tuning and their applications as Li-ion battery anode materials. Chemical Society Reviews. 2014;**43**(13):4341-4356. DOI: 10.1039/c4cs00071d

[28] Fujita M, Komatsu N, Kimura T. Sonochemical preparation of carbon spheres. Ultrasonics Sonochemistry. 2014;**21**(3):943-945. DOI: 10.1016/j.ultsonch.2013.11.013

[29] Kumar R, Kumar VB, Gedanken A. Sonochemical synthesis of carbon dots, mechanism, effect of parameters, and catalytic, energy, biomedical and tissue

engineering applications. Ultrasonics Sonochemistry. 2020;**64**:105009. DOI: 10.1016/j.ultsonch.2020.105009

[30] Kumar VB, Tang J, Lee KJ, Pol VG, Gedanken A. In situ sonochemical synthesis of luminescent Sn@C-dots and hybrid Sn@C-dots@Sn anode for lithium-ion batteries. RSC Advances. 2016;**6**(70):66256-66265. DOI: 10.1039/C6RA09926B

[31] Deosarkar MP, Pawar SM, Bhanvase BA. In situ sonochemical synthesis of Fe_3O_4-graphene nanocomposite for lithium rechargeable batteries. Chemical Engineering and Processing. 2014;**83**:49-55. DOI: 10.1016/j.cep.2014.07.004

[32] Wu K, Liu D, Lu W, Zhang K. One-pot sonochemical synthesis of magnetite@reduced graphene oxide nanocomposite for high performance Li ion storage. Ultrasonics Sonochemistry. 2018;**45**:167-172. DOI: 10.1016/j.ultsonch.2018.03.015

[33] Hamidi R, Ghasemi S, Hosseini SR. Ultrasonic assisted synthesis of $Ni_3(VO_4)_2$-reduced graphene oxide nanocomposite for potential use in electrochemical energy storage. Ultrasonics Sonochemistry. 2020;**62**:104869. DOI: 10.1016/j.ultsonch.2019.104869

[34] Chen Y, Chuang Y, Su JH, Yu HC, Chen-Yang YW. High discharge capacity solid composite polymer electrolyte lithium battery. Lancet. 2011;**196**:2802-2809. DOI: 10.1016/j.jpowsour.2010.11.058

[35] Nacimiento F, Alcantara R, Nwokeke UG, Gonzalez JR, Tirado JL. Nanocrystalline $CoSn_2$-carbon composite electrode prepared by using sonochemistry. Ultrasonics Sonochemistry. 2012;**19**(2):352-357. DOI: 10.1016/j.ultsonch.2011.06.014

[36] Xu C, Wang X, Yang L, Wu Y. Fabrication of a graphene-cuprous oxide composite. Journal of Solid State Chemistry. 2009;**182**:2486-2490. DOI: 10.1016/j.jssc.2009.07.001

[37] Abulizi A, Yang GH, Zhu JJ. One-step simple sonochemical fabrication and photocatalytic properties of Cu_2O-rGO composites. Ultrasonics Sonochemistry. 2014;**21**(1):129-135. DOI: 10.1016/j.ultsonch.2013.07.013

[38] Wang C, Higgins D, Wang F, Li D, Liu R, Xia G, et al. Controlled synthesis of micro/nanostructured CuO anodes for lithium-ion batteries. Nano Energy. 2014;**9**:334-344. DOI: 10.1016/j.nanoen.2014.08.009

[39] Park JC, Kim J, Kwon H, Song H. Gram-scale synthesis of Cu_2O nanocubes and subsequent oxidation to CuO hollow nanostructures for lithium-ion battery anode materials. Advanced Materials. 2009;**21**(7):803-807. DOI: 10.1002/adma.200800596

[40] Ko S, Lee JI, Yang HS, Park S, Jeong U. Mesoporous CuO particles threaded with CNTs for high-performance lithium-ion battery anodes. Advanced Materials. 2012;**24**(32):4451-4456. DOI: 10.1002/adma.201201821

[41] Mai YJ, Wang XL, Xiang JY, Qiao YQ, Zhang D, Gu CD, et al. CuO/graphene composite as anode materials for lithium-ion batteries. Electrochimica Acta. 2011;**56**(5):2306-2311. DOI: 10.1016/j.electacta.2010.11.036

[42] Hojaghani S, Hosaini Sadr M, Morsali A. Sonochemical synthesis of two new copper(II) complexes with azo ligands derived from anthranilic acid and beta-naphtol. Ultrasonics Sonochemistry. 2015;**26**:305-311. DOI: 10.1016/j.ultsonch.2015.02.009

[43] Parsaee Z, Joukar Bahaderani E, Afandak A. Sonochemical synthesis, in vitro evaluation and DFT study of novel phenothiazine base Schiff bases

and their nano copper complexes as the precursors for new shaped CuO-NPs. Ultrasonics Sonochemistry. 2018;**40**(Pt A):629-643. DOI: 10.1016/j.ultsonch.2017.08.010

[44] Alavi MA, Morsali A, Joo SW, Min BK. Ultrasound and modulation assisted synthesis of {[Cu_2(BDC-NH_2)$_2$(dabco)]DMF.3H_2O} nanostructures: New precursor to prepare nanorods and nanotubes of copper(II) oxide. Ultrasonics Sonochemistry. 2015;**22**:349-358. DOI: 10.1016/j.ultsonch.2014.04.017

[45] Hajnorouzi A. Two ultrasonic applications for the synthesis of nanostructured copper oxide (II). Ultrasonics Sonochemistry. 2020;**64**:105020. DOI: 10.1016/j.ultsonch.2020.105020

[46] Yang Z, Du Y, Hou G, Ouyang Y, Ding F, Yuan F. Nanoporous silicon spheres preparation via a controllable magnesiothermic reduction as anode for Li-ion batteries. Electrochimica Acta. 2020;**329**:135141. DOI: 10.1016/j.electacta.2019.135141

[47] Ge M, Rong J, Fang X, Zhang A, Lu Y, Zhou C. Scalable preparation of porous silicon nanoparticles and their application for lithium-ion battery anodes. Nano Research. 2013;**6**(3):174-181. DOI: 10.1007/s12274-013-0293-y

[48] Zhao Y, Liu X, Li H, Zhai T, Zhou H. Hierarchical micro/nanoporous silicon Li-ion battery anodes. Chemical Communications. 2012;**48**(42):5079-5081. DOI: 10.1039/c2cc31476b

[49] Cadiz Bedini AP, Klingebiel B, Luysberg M, Carius R. Sonochemical synthesis of hydrogenated amorphous silicon nanoparticles from liquid trisilane at ambient temperature and pressure. Ultrasonics Sonochemistry. 2017;**39**:883-888. DOI: 10.1016/j.ultsonch.2017.06.011

[50] Li XL, Li TC, Huang S, Zhang J, Pam ME, Yang HY. Controllable synthesis of two-dimensional molybdenum disulfide (MoS_2) for energy-storage applications. ChemSusChem. 2020;**13**(6):1379-1391. DOI: 10.1002/cssc.201902706

[51] Xu W, Sun C, Zhao K, Cheng X, Rawal S, Xu Y, et al. Defect engineering activating (boosting) zinc storage capacity of MoS_2. Energy Storage Materials. 2019;**16**:527-534. DOI: 10.1016/j.ensm.2018.09.009

[52] Liang Y, Feng R, Yang S, Ma H, Liang J, Chen J. Rechargeable Mg batteries with graphene-like MoS(2) cathode and ultrasmall Mg nanoparticle anode. Advanced Materials. 2011;**23**(5):640-643. DOI: 10.1002/adma.201003560

[53] Liu Y, Li R. Study on ultrasound-assisted liquid-phase exfoliation for preparing graphene-like molybdenum disulfide nanosheets. Ultrasonics Sonochemistry. 2020;**63**:104923. DOI: 10.1016/j.ultsonch.2019.104923

[54] Li Y, Yao J, Uchaker E, Yang J, Huang Y, Zhang M, et al. Leaf-like V_2O_5 nanosheets fabricated by a facile green approach as high energy cathode material for lithium-ion batteries. Advanced Energy Materials. 2013;**3**(9):1171-1175. DOI: 10.1002/aenm.201300188

[55] Kumar VG, Kim KB. Organized and highly dispersed growth of MnO_2 nano-rods by sonochemical hydrolysis of Mn_3 acetate. Ultrasonics Sonochemistry. 2006;**13**(6):549-556. DOI: 10.1016/j.ultsonch.2005.09.010

[56] Okitsu K, Iwatani M, Nanzai B, Nishimura R, Maeda Y. Sonochemical reduction of permanganate to manganese dioxide: The effects of H_2O_2 formed in the sonolysis of water on the rates of reduction. Ultrasonics Sonochemistry. 2009;**16**(3):387-391. DOI: 10.1016/j.ultsonch.2008.10.009

[57] Gnana Sundara Raj B, Asiri AM, Qusti AH, Wu JJ, Anandan S. Sonochemically synthesized MnO_2 nanoparticles as electrode material for supercapacitors. Ultrasonics Sonochemistry. 2014;**21**(6):1933-1938. DOI: 10.1016/j.ultsonch.2013.11.018

[58] Ganesh Kumar V, Aurbuch D, Gedanken A. A comparison between hot-hydrolysis and sonolysis of various Mn(II) salts. Ultrasonics Sonochemistry. 2003;**10**(1):17-23. DOI: 10.1016/S1350-4177(02)00091-3

[59] Sivakumar P, Nayak PK, Markovsky B, Aurbach D, Gedanken A. Sonochemical synthesis of $LiNi_{0.5}Mn_{1.5}O_4$ and its electrochemical performance as a cathode material for 5 V Li-ion batteries. Ultrasonics Sonochemistry. 2015;**26**:332-339. DOI: 10.1016/j.ultsonch.2015.02.007

[60] Fan X, Liu J, Song Z, Han X, Deng Y, Zhong C, et al. Porous nanocomposite gel polymer electrolyte with high ionic conductivity and superior electrolyte retention capability for long-cycle-life flexible zinc-air batteries. Nano Energy. 2019;**56**:454-462. DOI: 10.1016/j.nanoen.2018.11.057

[61] Zhu L, Zhang H. A novel method for the modification of zinc powder by ultrasonic impregnation in cerium nitrate solution. Ultrasonics Sonochemistry. 2008;**15**:393-401. DOI: 10.1016/j.ultsonch.2007.09.016

[62] Chen L, Zhang Y, Dong L, Liu X, Long L, Wang S, et al. Honeycomb-like 3D N-, P-codoped porous carbon anchored with ultrasmall Fe_2P nanocrystals for efficient Zn-air battery. Carbon. 2020;**158**:885-892. DOI: 10.1016/j.carbon.2019.11.073

[63] Chen T-W, Chinnapaiyan S, Chen S-M, Ali MA, Elshikh MS, Lee S-Y, et al. Sonochemical approach to the synthesis of metal tungstate/nafion composite with electrocatalytic properties and its electrochemical sensing performance. Ultrasonics Sonochemistry. 2020;**66**:104901. DOI: 10.1016/j.ultsonch.2019.104901

[64] Munonde TS, Zheng H, Nomngongo PN. Ultrasonic exfoliation of NiFe LDH/CB nanosheets for enhanced oxygen evolution catalysis. Ultrasonics Sonochemistry. 2019;**59**:104716. DOI: 10.1016/j.ultsonch.2019.104716

[65] Xie J, Cao X, Li J, Zhan H, Xia Y, Zhou Y. Application of ultrasonic irradiation to the sol-gel synthesis of silver vanadium oxides. Ultrasonics Sonochemistry. 2005;**12**(4):289-293. DOI: 10.1016/j.ultsonch.2004.01.041

[66] Park J, Yu S-H, Sung Y-E. Design of structural and functional nanomaterials for lithium-sulfur batteries. Nano Today. 2018;**18**:35-64. DOI: 10.1016/j.nantod.2017.12.010

[67] Li Z, Jiang Y, Yuan L, Yi Z, Wu C, Liu Y, et al. A highly ordered meso@micro-porous carbon supported sulfur@smaller-sulfur core-shell structured cathode for Li-S batteries. ACS Nano. 2014;**8**(9):9295-9303. DOI: 10.1021/nn503220h

[68] Bagotsky VS, Skundin AM, Volfkovich MY. Electrochemical Power Sources: Batteries, Fuel Cells, and Supercapacitors. Hoboken: Wiley; 2015. DOI: 10.1002/9781118942857

[69] Sealy C. The problem with platinum. Materials Today. 2008;**11**(12):65-68. DOI: 10.1016/s1369-7021(08)70254-2

[70] Karousos DS, Desdenakis KI, Sakkas PM, Sourkouni G, Pollet BG, Argirusis C. Sonoelectrochemical one-pot synthesis of Pt-carbon black nanocomposite PEMFC electrocatalyst. Ultrasonics Sonochemistry. 2017;**35**(Pt B):591-597. DOI: 10.1016/j.ultsonch.2016.05.023

[71] Gong K, Du F, Xia Z, Durstock M, Dai L. Nitrogen-doped carbon nanotube arrays with high electrocatalytic activity for oxygen reduction. Science. 2009;**323**(5915):760-764. DOI: 10.1126/science.1168049

[72] Mentus S, Abu Rabi A, Jašin D. Oxygen reduction on potentiodynamically formed Pd/TiO_2 composite electrodes. Electrochimica Acta. 2012;**69**:174-180. DOI: 10.1016/j.electacta.2012.02.106

[73] Truong-Phuoc L, Pham-Huu C, Da Costa V, Janowska I. Few-layered graphene-supported palladium as a highly efficient catalyst in oxygen reduction reaction. Chemical Communications. 2014;**50**(92): 14433-14435. DOI: 10.1039/c4cc05527f

[74] Padilla RH, Priecel P, Lin M, Lopez-Sanchez JA, Zhong Z. A versatile sonication-assisted deposition-reduction method for preparing supported metal catalysts for catalytic applications. Ultrasonics Sonochemistry. 2017;**35**(Pt B):631-639. DOI: 10.1016/j.ultsonch.2016.01.018

[75] Lee E, Kwon YU. Multi-component electrocatalyst for low-temperature fuel cells synthesized via sonochemical reactions. Ultrasonics Sonochemistry. 2016;**29**:401-412. DOI: 10.1016/j.ultsonch.2015.10.013

[76] Lee E, Jang JH, Matin MA, Kwon YU. One-step sonochemical syntheses of Ni@Pt core-shell nanoparticles with controlled shape and shell thickness for fuel cell electrocatalyst. Ultrasonics Sonochemistry. 2014;**21**(1):317-323. DOI: 10.1016/j.ultsonch.2013.05.006

[77] Park HU, Park AH, Shi W, Park GG, Kwon YU. Ternary core-shell PdM@Pt (M=Mn and Fe) nanoparticle electrocatalysts with enhanced ORR catalytic properties. Ultrasonics Sonochemistry. 2019;**58**:104673. DOI: 10.1016/j.ultsonch.2019.104673

[78] Anandan S, Wu JJ. Sonochemical synthesis of carbon supported Sn nanoparticles and its electrochemical application. Ultrasonics Sonochemistry. 2014;**21**(6):1954-1957. DOI: 10.1016/j.ultsonch.2014.02.018

[79] Zheng H, Matseke MS, Munonde TS. The unique Pd@Pt/C core-shell nanoparticles as methanol-tolerant catalysts using sonochemical synthesis. Ultrasonics Sonochemistry. 2019;**57**:166-171. DOI: 10.1016/j.ultsonch.2019.05.023

[80] Wang H, Zhang W, Bai P, Xu L. Ultrasound-assisted transformation from waste biomass to efficient carbon-based metal-free pH-universal oxygen reduction reaction electrocatalysts. Ultrasonics Sonochemistry. 2020;**65**:105048. DOI: 10.1016/j.ultsonch.2020.105048

[81] Zuo LX, Jiang LP, Abdel-Halim ES, Zhu JJ. Sonochemical preparation of stable porous MnO_2 and its application as an efficient electrocatalyst for oxygen reduction reaction. Ultrasonics Sonochemistry. 2017;**35**(Pt A):219-225. DOI: 10.1016/j.ultsonch.2016.09.021

[82] Esmaielzadeh S, Ahmadizadegan H. Construction of proton exchange membranes under ultrasonic irradiation based on novel fluorine functionalizing sulfonated polybenzimidazole/cellulose/silica bionanocomposite. Ultrasonics Sonochemistry. 2018;**41**:641-650. DOI: 10.1016/j.ultsonch.2017.10.029

[83] Zuo LX, Jiang LP, Zhu JJ. A facile sonochemical route for the synthesis of MoS_2/Pd composites for highly efficient oxygen reduction reaction. Ultrasonics Sonochemistry. 2017;**35**(Pt B):681-688. DOI: 10.1016/j.ultsonch.2016.02.006

[84] Hoogers G. Chapter 4: Fuel cell components and their impact on performance. In: Hoogers G, editor. Fuel Cell Technology Handbook. Boca Raton: CRC Press; 2002. pp. 88-114. DOI: 10.1201/9781420041552

[85] Carrette L, Friedrich KA, Stimming U. Fuel cells – Fundamentals and applications. Fuel Cells. 2001;**1**(1):5-39. DOI: 10.1002/1615-6854(200105)1:1<5::aid-fuce5>3.0.co;2-g

[86] Islam MH, Burheim OS, Pollet BG. Sonochemical and sonoelectrochemical production of hydrogen. Ultrasonics Sonochemistry. 2019;**51**:533-555. DOI: 10.1016/j.ultsonch.2018.08.024

[87] Cherepanov PV, Ashokkumar M, Andreeva DV. Ultrasound assisted formation of Al-Ni electrocatalyst for hydrogen evolution. Ultrasonics Sonochemistry. 2015;**23**:142-147. DOI: 10.1016/j.ultsonch.2014.10.012

[88] Ida S, Wilson P, Neppolian B, Sathish M, Mahammed Shaheer AR, Ravi P. Tuning the type of nitrogen on N-RGO supported on N-TiO_2 under ultrasonication/hydrothermal treatment for efficient hydrogen evolution - A mechanistic overview. Ultrasonics Sonochemistry. 2020;**64**:104866. DOI: 10.1016/j.ultsonch.2019.104866

[89] Steele B. Fuel-cell technology-running on natural gas. Nature. 1999;**400**:619-621. DOI: 10.1038/23144

[90] Argirusis C, Damjanović T, Borchardt G. Electrophoretic deposition of thin SOFC-electrolyte films on porous $La_{0.75}Sr_{0.2}MnO_{3-\delta}$ cathodes. Materials Science Forum. 2004;**453-454**:335-342. DOI: 10.4028/www.scientific.net/MSF.453-454.335

[91] Argirusis C, Damjanović T, Borchardt G. Preparation of SOFC cells by means of electrophoretic deposition. Key Engineering Materials. 2006;**314**:101-106. DOI: 10.4028/www.scientific.net/KEM.314.101

[92] Argirusis C, Matić S, Schneider O. An EQCM study of ultrasonically assisted electrodeposition of Co/CeO_2 and Ni/CeO_2 composites for fuel cell applications. Physica Status Solidi A: Applications and Materials Science. 2008;**205**(10):2400-2404. DOI: 10.1002/pssa.200779409

[93] Argirusis C, Grosse-Brauckmann J, Sourkouni G, Taillades G, Roziere J. Preparation of thin proton conducting membranes by means of EPD. Key Engineering Materials. 2009;**412**: 125-130. DOI: 10.4028/www.scientific.net/KEM.412.125

[94] Szepanski C, Grosse-Brauckmann J, Argirusis C. Electrophoretic deposition as preparation method for intermediate temperature SOFC half cells. Key Engineering Materials. 2009;**412**: 209-214. DOI: 10.4028/www.scientific.net/KEM.412.209

[95] Kharlamova T, Pavlova S, Sadykov V, Krieger T, Batuev L, Muzykantov V, et al. Fe- and Al-doped apatite-type lanthanum silicates: Structure and property characterization. Solid State Ionics. 2009;**180**(11):796-799. DOI: 10.1016/j.ssi.2008.12.042

[96] Jothinathan E, Vanmeensel K, Vleugels J, Kharlamova T, Sadykov V, Pavlova S, et al. Apatite type lanthanum silicate and composite anode half cells. Solid State Ionics. 2011;**192**(1):419-423. DOI: 10.1016/j.ssi.2010.02.009

[97] Kharlamova T, Pavlova S, Sadykov V, Krieger T, Alikina G, Frade J, et al. Anode composites based on NiO and apatite-type lanthanum silicate for intermediate temperature solid oxide fuel cells. MRS Proceedings. 2008;**1098**:104-109. DOI: 10.1557/PROC-1098-HH07-02

[98] Kharlamova T, Pavlova S, Sadykov V, Krieger T, Mezentseva N, Muzykantov V, et al. Perovskite and composite materials for intermediate temperatures solid oxide fuel cells. MRS Proceedings. 2008;**1056**:102-107. DOI: 10.1557/PROC-1056-HH03-64

[99] Kharlamova T, Pavlova S, Sadykov V, Krieger T, Alikina G, Argirusis C. Catalytic properties and coking stability of new anode materials for internal methane reforming in the intermediate temperature solid oxide fuel cells. Catalysis Today. 2009;**146**(1-2):141-147. DOI: 10.1016/j.cattod.2009.01.052

[100] Jiang S, Chan SH. A review of anode materials development in solid oxide fuel cells. Journal of Materials Science. 2004;**39**:4405-4439. DOI: 10.1023/B:JMSC.0000034135.52164.6b

[101] Sakkas PM, Bozes CY, Kanellopoulou DG, Sourkouni G, Argirusis C. A study on the synchronous decoration of molybdenum oxide or tungsten oxide nanoparticles on anode materials for natural gas fed solid oxide fuel cells using ultrasounds. Ultrasonics Sonochemistry. 2019;**59**:104715. DOI: 10.1016/j.ultsonch.2019.104715

[102] Okkay H, Bayramoglu M, Oksuzomer MF. $Ce_{0.8}Sm_{0.2}O_{1.9}$ synthesis for solid oxide fuel cell electrolyte by ultrasound assisted co-precipitation method. Ultrasonics Sonochemistry. 2013;**20**(3):978-983. DOI: 10.1016/j.ultsonch.2012.10.013

[103] Pinjari DV, Pandit AB. Room temperature synthesis of crystalline CeO_2 nanopowder: Advantage of sonochemical method over conventional method. Ultrasonics Sonochemistry. 2011;**18**(5):1118-1123. DOI: 10.1016/j.ultsonch.2011.01.008

[104] Gohari Derakhshandeh P, Soleimannejad J. Sonochemical synthesis of a new nano-sized cerium(III) supramolecular compound; precursor for nanoceria. Ultrasonics Sonochemistry. 2016;**31**:122-128. DOI: 10.1016/j.ultsonch.2015.12.014

[105] Yang G, Zhou Y, Pan HB, Zhu C, Fu S, Wai CM, et al. Ultrasonic-assisted synthesis of Pd-Pt/carbon nanotubes nanocomposites for enhanced electro-oxidation of ethanol and methanol in alkaline medium. Ultrasonics Sonochemistry. 2016;**28**:192-198. DOI: 10.1016/j.ultsonch.2015.07.021

[106] Douk AS, Saravani H, Farsadrooh M, Noroozifar M. An environmentally friendly one-pot synthesis method by the ultrasound assistance for the decoration of ultrasmall Pd-Ag NPs on graphene as high active anode catalyst towards ethanol oxidation. Ultrasonics Sonochemistry. 2019;**58**:104616. DOI: 10.1016/j.ultsonch.2019.104616

[107] Farsadrooh M, Noroozifar M, Modarresi-Alam AR, Saravani H. Sonochemical synthesis of high-performance Pd@CuNWs/MWCNTs-CH electrocatalyst by galvanic replacement toward ethanol oxidation in alkaline media. Ultrasonics Sonochemistry. 2019;**51**:478-486. DOI: 10.1016/j.ultsonch.2018.06.011

[108] Sakkas PM, Schneider O, Sourkouni G, Argirusis C. Sonochemistry in the service of SOFC research. Ultrasonics Sonochemistry. 2014;**21**(6):1939-1947. DOI: 10.1016/j.ultsonch.2014.02.001

[109] Sakkas P, Schneider O, Martens S, Thanou P, Sourkouni G, Argirusis C. Fundamental studies of sonoelectrochemical nanomaterials preparation. Journal of Applied Electrochemistry. 2012;**42**(9):763-777. DOI: 10.1007/s10800-012-0443-z

[110] Islam MH, Paul MTY, Burheim OS, Pollet BG. Recent developments in the sonoelectrochemical synthesis of nanomaterials. Ultrasonics Sonochemistry. 2019;**59**:104711. DOI: 10.1016/j.ultsonch.2019.104711

[111] Colomer MT, Anderson MA. High porosity silica xerogels prepared by a particulate sol-gel route: Pore structure and proton conductivity. Journal of Non-Crystalline Solids.

2001;**290**(2):93-104. DOI: 10.1016/S0022-3093(01)00815-8

[112] Cortes AD, Donatti DA, Ibanez Ruiz A, Vollet DR. A kinetic study of the effect of ultrasound power on the sonohydrolysis of tetraethyl orthosilicate. Ultrasonics Sonochemistry. 2007;**14**(6):711-716. DOI: 10.1016/j.ultsonch.2006.12.002

[113] Wu C, Liu L, Tang K, Chen T. Studies on an ultrasonic atomization feed direct methanol fuel cell. Ultrasonics Sonochemistry. 2017;**34**:60-66. DOI: 10.1016/j.ultsonch.2016.05.018

[114] Wu C, Wu J, Luo H, Wang S, Chen T. Ultrasonic radiation to enable improvement of direct methanol fuel cell. Ultrasonics Sonochemistry. 2016;**29**:363-370. DOI: 10.1016/j.ultsonch.2015.10.012

[115] Hunge YM. Basics and advanced developments in photocatalysis – A review (mini review). International Journal of Hydrology. 2018;**2**(4):539-540. DOI: 10.15406/ijh.2018.02.00122

[116] Serpone N, Emeline AV. Semiconductor photocatalysis - past, present, and future outlook. Journal of Physical Chemistry Letters. 2012;**3**(5):673-677. DOI: 10.1021/jz300071j

[117] Stucchi M, Bianchi CL, Argirusis C, Pifferi V, Neppolian B, Cerrato G, et al. Ultrasound assisted synthesis of Ag-decorated TiO_2 active in visible light. Ultrasonics Sonochemistry. 2018;**40**(Pt A):282-288. DOI: 10.1016/j.ultsonch.2017.07.016

[118] Stucchi M, Bianchi CL, Pirola C, Cerrato G, Morandi S, Argirusis C, et al. Copper NPs decorated titania: A novel synthesis by high energy US with a study of the photocatalytic activity under visible light. Ultrasonics Sonochemistry. 2016;**31**:295-301. DOI: 10.1016/j.ultsonch.2016.01.015

[119] Cheng C, Amini A, Zhu C, Xu Z, Song H, Wang N. Enhanced photocatalytic performance of TiO_2-ZnO hybrid nanostructures. Scientific Reports. 2014;**4**:4181. DOI: 10.1038/srep04181

[120] Hunge YM, Yadav AA, Mathe VL. Ultrasound assisted synthesis of WO_3-ZnO nanocomposites for brilliant blue dye degradation. Ultrasonics Sonochemistry. 2018;**45**:116-122. DOI: 10.1016/j.ultsonch.2018.02.052

[121] Mohite VS, Mahadik MA, Kumbhar SS, Hunge YM, Kim JH, Moholkar AV, et al. Photoelectrocatalytic degradation of benzoic acid using Au doped TiO_2 thin films. Journal of Photochemistry and Photobiology. B. 2015;**142**:204-211. DOI: 10.1016/j.jphotobiol.2014.12.004

[122] Hunge YM, Mahadik MA, Moholkar AV, Bhosale CH. Photoelectrocatalytic degradation of oxalic acid using WO_3 and stratified WO_3/TiO_2 photocatalysts under sunlight illumination. Ultrasonics Sonochemistry. 2017;**35**(Pt A):233-242. DOI: 10.1016/j.ultsonch.2016.09.024

[123] Vallejo W, Rueda A, Diaz-Uribe C, Grande C, Quintana P. Photocatalytic activity of graphene oxide-TiO_2 thin films sensitized by natural dyes extracted from *Bactris guineensis*. Royal Society Open Science. 2019;**6**(3):181824. DOI: 10.1098/rsos.181824

[124] Xie H, Liu B, Zhao X. Facile process to greatly improve the photocatalytic activity of the TiO_2 thin film on window glass for the photodegradation of acetone and benzene. Chemical Engineering Journal. 2016;**284**:1156-1164. DOI: 10.1016/j.cej.2015.09.049

[125] Jiang L, Wang Y, Feng C. Application of photocatalytic technology in environmental safety. Procedia Engineering. 2012;**45**:993-997. DOI: 10.1016/j.proeng.2012.08.271

[126] Qin Z, Su T, Ji H. Photocatalytic nanomaterials for the energy and environmental application. In: Multifunctional Nanocomposites for Energy and Environmental Applications. Vol. 1. New York: Wiley; 2018. pp. 353-401. DOI: 10.1002/9783527342501.ch13

[127] Muhammad NA, Wang Y, Muhammad FE, He T. Photoreduction of carbon dioxide using strontium zirconate nanoparticles. Science China Materials. 2015;**58**(8):634-639. DOI: 10.1007/s40843-015-0077-7

[128] Akhter P, Hussain M, Saracco G, Russo N. Novel nanostructured-TiO_2 materials for the photocatalytic reduction of CO_2 greenhouse gas to hydrocarbons and syngas. Fuel. 2015;**149**:55-65. DOI: 10.1016/j.fuel.2014.09.079

[129] Chen J, Xin F, Yin X, Xiang T, Wang Y. Synthesis of hexagonal and cubic $ZnIn_2S_4$ nanosheets for the photocatalytic reduction of CO_2 with methanol. RSC Advances. 2015;**5**(5):3833-3839. DOI: 10.1039/c4ra13191f

[130] Ghows N, Entezari MH. Sono-synthesis of core-shell nanocrystal (CdS/TiO_2) without surfactant. Ultrasonics Sonochemistry. 2012;**19**(5):1070-1078. DOI: 10.1016/j.ultsonch.2012.01.009

[131] Mousavi-Kamazani M, Salavati-Niasari M, Emadi H. Preparation of stochiometric $CuInS_2$ nanostructures by ultrasonic method. Micro & Nano Letters. 2012;**7**(9):896-900. DOI: 10.1049/mnl.2012.0393

[132] Kolny-Olesiak J, Weller H. Synthesis and application of colloidal $CuInS_2$ semiconductor nanocrystals. ACS Applied Materials & Interfaces. 2013;**5**(23): 12221-12237. DOI: 10.1021/am404084d

[133] Kansaard T, Pecharapa W. Characterization of $BiVO_4$ nanoparticles prepared by sonochemical process. Ferroelectrics. 2019;**552**(1):140-147. DOI: 10.1080/00150193.2019.1653090

[134] Liu S, Zhang N, Tang ZR, Xu YJ. Synthesis of one-dimensional CdS@TiO(2) core-shell nanocomposites photocatalyst for selective redox: The dual role of TiO(2) shell. ACS Applied Materials & Interfaces. 2012;**4**(11): 6378-6385. DOI: 10.1021/am302074p

[135] Koltypin Y, Nikitenko SI, Gedanken A. The sonochemical preparation of tungsten oxide nanoparticles. Journal of Materials Chemistry. 2002;**12**(4):1107-1110. DOI: 10.1039/b106036h

[136] Wang L, Wang Y, Cheng Y, Liu Z, Guo Q, Ha MN, et al. Hydrogen-treated mesoporous WO_3 as a reducing agent of CO_2 to fuels (CH_4 and CH_3OH) with enhanced photothermal catalytic performance. Journal of Materials Chemistry A. 2016;**4**(14):5314-5322. DOI: 10.1039/c5ta10180h

[137] Jin J, Yu J, Guo D, Cui C, Ho W. A hierarchical Z-scheme CdS-WO_3 photocatalyst with enhanced CO_2 reduction activity. Small. 2015;**11**(39):5262-5271. DOI: 10.1002/smll.201500926

[138] Dong Y-z, Xue Y-s, Yang W-w, You H-m, Su Y. Visible light driven CdS/WO_3 inverse opals with enhanced RhB degradation activity. Colloids and Surfaces A. 2019;**561**:381-387. DOI: 10.1016/j.colsurfa.2018.10.033

[139] Wang S, Guan BY, Lou XWD. Construction of $ZnIn_2S_4$-In_2O_3 hierarchical tubular heterostructures for efficient CO_2 photoreduction. Journal of the American Chemical Society. 2018;**140**(15):5037-5040. DOI: 10.1021/jacs.8b02200

[140] Yuan J, Hao C. Solar-driven photoelectrochemical reduction of carbon dioxide to methanol at $CuInS_2$

thin film photocathode. Solar Energy Materials and Solar Cells. 2013;**108**:170-174. DOI: 10.1016/j.solmat.2012.09.024

[141] Paz FA, Klinowski J, Vilela SM, Tome JP, Cavaleiro JA, Rocha J. Ligand design for functional metal-organic frameworks. Chemical Society Reviews. 2012;**41**(3):1088-1110. DOI: 10.1039/c1cs15055c

[142] Yaghi OM, O'Keeffe M, Ockwig NW, Chae HK, Eddaoudi M, Kim J. Reticular synthesis and the design of new materials. Nature. 2003;**423**(6941):705-714. DOI: 10.1038/nature01650

[143] Zhao D, Timmons DJ, Yuan D, Zhou HC. Tuning the topology and functionality of metal-organic frameworks by ligand design. Accounts of Chemical Research. 2011;**44**(2):123-133. DOI: 10.1021/ar100112y

[144] Ma S, Zhou HC. A metal-organic framework with entatic metal centers exhibiting high gas adsorption affinity. Journal of the American Chemical Society. 2006;**128**(36):11734-11735. DOI: 10.1021/ja063538z

[145] González-Zamora E, Ibarra IA. CO_2 capture under humid conditions in metal-organic frameworks. Materials Chemistry Frontiers. 2017;**1**(8):1471-1484. DOI: 10.1039/c6qm00301j

[146] Li JR, Sculley J, Zhou HC. Metal-organic frameworks for separations. Chemical Reviews. 2012;**112**(2):869-932. DOI: 10.1021/cr200190s

[147] Liu J, Chen L, Cui H, Zhang J, Zhang L, Su CY. Applications of metal-organic frameworks in heterogeneous supramolecular catalysis. Chemical Society Reviews. 2014;**43**(16):6011-6061. DOI: 10.1039/c4cs00094c

[148] Zhu Q, Chen Y, Wang W, Zhang H, Ren C, Chen H, et al. A sensitive biosensor for dopamine determination based on the unique catalytic chemiluminescence of metal-organic framework HKUST-1. Sensors and Actuators B: Chemical. 2015;**210**:500-507. DOI: 10.1016/j.snb.2015.01.012

[149] Bobbitt NS, Mendonca ML, Howarth AJ, Islamoglu T, Hupp JT, Farha OK, et al. Metal-organic frameworks for the removal of toxic industrial chemicals and chemical warfare agents. Chemical Society Reviews. 2017;**46**(11):3357-3385. DOI: 10.1039/c7cs00108h

[150] Hu Z, Deibert BJ, Li J. Luminescent metal-organic frameworks for chemical sensing and explosive detection. Chemical Society Reviews. 2014;**43**(16):5815-5840. DOI: 10.1039/c4cs00010b

[151] Li X, Liu Y, Wang J, Gascon J, Li J, Van der Bruggen B. Metal-organic frameworks based membranes for liquid separation. Chemical Society Reviews. 2017;**46**(23):7124-7144. DOI: 10.1039/c7cs00575j

[152] Abazari R, Reza Mahjoub A, Slawin AMZ, Carpenter-Warren CL. Morphology- and size-controlled synthesis of a metal-organic framework under ultrasound irradiation: An efficient carrier for pH responsive release of anti-cancer drugs and their applicability for adsorption of amoxicillin from aqueous solution. Ultrasonics Sonochemistry. 2018;**42**:594-608. DOI: 10.1016/j.ultsonch.2017.12.032

[153] Fang Y, Ma Y, Zheng M, Yang P, Asiri AM, Wang X. Metal-organic frameworks for solar energy conversion by photoredox catalysis. Coordination Chemistry Reviews. 2018;**373**:83-115. DOI: 10.1016/j.ccr.2017.09.013

[154] Rajpurohit AS, Punde NS, Srivastava AK. A dual metal organic framework based on copper-iron clusters integrated sulphur doped

graphene as a porous material for supercapacitor with remarkable performance characteristics. Journal of Colloid and Interface Science. 2019;**553**:328-340. DOI: 10.1016/j.jcis.2019.06.031

[155] Vayenas M, Vaitsis C, Sourkouni G, Pandis PK, Argirusis C. Investigation of alternative materials as bifunctional catalysts for electrochemical applications. Chimica Techno Acta. 2019;**6**(4):120-129. DOI: 10.15826/chimtech.2019.6.4.01

[156] Ren Y, Chia GH, Gao Z. Metal-organic frameworks in fuel cell technologies. Nano Today. 2013;**8**(6):577-597. DOI: 10.1016/j.nantod.2013.11.004

[157] Tranchemontagne DJ, Hunt JR, Yaghi OM. Room temperature synthesis of metal-organic frameworks: MOF-5, MOF-74, MOF-177, MOF-199, and IRMOF-0. Tetrahedron. 2008;**64**(36):8553-8557. DOI: 10.1016/j.tet.2008.06.036

[158] Choi J-S, Son W-J, Kim J, Ahn W-S. Metal-organic framework MOF-5 prepared by microwave heating: Factors to be considered. Microporous and Mesoporous Materials. 2008;**116**(1-3):727-731. DOI: 10.1016/j.micromeso.2008.04.033

[159] Wu X, Bao Z, Yuan B, Wang J, Sun Y, Luo H, et al. Microwave synthesis and characterization of MOF-74 (M=Ni, Mg) for gas separation. Microporous and Mesoporous Materials. 2013;**180**:114-122. DOI: 10.1016/j.micromeso.2013.06.023

[160] Martinez Joaristi A, Juan-Alcañiz J, Serra-Crespo P, Kapteijn F, Gascon J. Electrochemical synthesis of some archetypical Zn^{2+}, Cu^{2+}, and Al^{3+} metal organic frameworks. Crystal Growth & Design. 2012;**12**(7):3489-3498. DOI: 10.1021/cg300552w

[161] Klimakow M, Klobes P, Thünemann AF, Rademann K, Emmerling F. Mechanochemical synthesis of metal-organic frameworks: A fast and facile approach toward quantitative yields and high specific surface areas. Chemistry of Materials. 2010;**22**(18):5216-5221. DOI: 10.1021/cm1012119

[162] Kleist W, Maciejewski M, Baiker A. MOF-5 based mixed-linker metal-organic frameworks: Synthesis, thermal stability and catalytic application. Thermochimica Acta. 2010;**499**(1-2):71-78. DOI: 10.1016/j.tca.2009.11.004

[163] Sun W, Zhai X, Zhao L. Synthesis of ZIF-8 and ZIF-67 nanocrystals with well-controllable size distribution through reverse microemulsions. Chemical Engineering Journal. 2016;**289**:59-64. DOI: 10.1016/j.cej.2015.12.076

[164] da Silva GG, Silva CS, Ribeiro RT, Ronconi CM, Barros BS, Neves JL, et al. Sonoelectrochemical synthesis of metal-organic frameworks. Synthetic Metals. 2016;**220**:369-373. DOI: 10.1016/j.synthmet.2016.07.003

[165] Qiu LG, Li ZQ, Wu Y, Wang W, Xu T, Jiang X. Facile synthesis of nanocrystals of a microporous metal-organic framework by an ultrasonic method and selective sensing of organoamines. Chemical Communications. 2008;**31**:3642-3644. DOI: 10.1039/b804126a

[166] Son WJ, Kim J, Kim J, Ahn WS. Sonochemical synthesis of MOF-5. Chemical Communications. 2008;**47**:6336-6338. DOI: 10.1039/b814740j

[167] Li Z-Q, Qiu L-G, Wang W, Xu T, Wu Y, Jiang X. Fabrication of nanosheets of a fluorescent metal-organic framework [Zn(BDC)(H_2O)]n (BDC=1,4-benzenedicarboxylate): Ultrasonic synthesis and sensing of ethylamine. Inorganic Chemistry

Communications. 2008;**11**(11):1375-1377. DOI: 10.1016/j.inoche.2008.09.010

[168] Haque E, Khan NA, Park JH, Jhung SH. Synthesis of a metal-organic framework material, iron terephthalate, by ultrasound, microwave, and conventional electric heating: A kinetic study. Chemistry - A European Journal. 2010;**16**(3):1046-1052. DOI: 10.1002/chem.200902382

[169] Vaitsis C, Sourkouni G, Argirusis C. Metal organic frameworks (MOFs) and ultrasound: A review. Ultrasonics Sonochemistry. 2019;**52**:106-119. DOI: 10.1016/j.ultsonch.2018.11.004

[170] Vaitsis C, Sourkouni G, Argirusis C. Sonochemical synthesis of MOFs. In: Mozafari M, editor. Metal-Organic Frameworks for Biomedical Applications. Cambridge: Woodhead Publishing; 2020. pp. 223-244. DOI: 10.1016/B978-0-12-816984-1.00013-5

[171] Zorpas AA. Strategy development in the framework of waste management. Science of the Total Environment. 2020;**716**:137088. DOI: 10.1016/j.scitotenv.2020.137088

[172] Loizia P, Neofytou N, Zorpas AA. The concept of circular economy strategy in food waste management for the optimization of energy production through anaerobic digestion. Environmental Science and Pollution Research International. 2019;**26**(15):14766-14773. DOI: 10.1007/s11356-018-3519-4

[173] Abbasi AR, Karimi M, Daasbjerg K. Efficient removal of crystal violet and methylene blue from wastewater by ultrasound nanoparticles Cu-MOF in comparison with mechanosynthesis method. Ultrasonics Sonochemistry. 2017;**37**:182-191. DOI: 10.1016/j.ultsonch.2017.01.007

[174] Tanhaei M, Mahjoub AR, Safarifard V. Sonochemical synthesis of amide-functionalized metal-organic framework/graphene oxide nanocomposite for the adsorption of methylene blue from aqueous solution. Ultrasonics Sonochemistry. 2018;**41**:189-195. DOI: 10.1016/j.ultsonch.2017.09.030

[175] Zhao S, Chen D, Wei F, Chen N, Liang Z, Luo Y. Removal of Congo red dye from aqueous solution with nickel-based metal-organic framework/graphene oxide composites prepared by ultrasonic wave-assisted ball milling. Ultrasonics Sonochemistry. 2017;**39**:845-852. DOI: 10.1016/j.ultsonch.2017.06.013

[176] Abazari R, Salehi G, Mahjoub AR. Ultrasound-assisted preparation of a nanostructured zinc(II) amine pillar metal-organic framework as a potential sorbent for 2,4-dichlorophenol adsorption from aqueous solution. Ultrasonics Sonochemistry. 2018;**46**:59-67. DOI: 10.1016/j.ultsonch.2018.02.001

[177] Abazari R, Mahjoub AR. Ultrasound-assisted synthesis of zinc(II)-based metal organic framework nanoparticles in the presence of modulator for adsorption enhancement of 2,4-dichlorophenol and amoxicillin. Ultrasonics Sonochemistry. 2018;**42**:577-584. DOI: 10.1016/j.ultsonch.2017.12.027

[178] Abbasi AR, Rizvandi M. Influence of the ultrasound-assisted synthesis of Cu-BTC metal-organic frameworks nanoparticles on uptake and release properties of rifampicin. Ultrasonics Sonochemistry. 2018;**40**(Pt A):465-471. DOI: 10.1016/j.ultsonch.2017.07.041

[179] Abbasi AR, Rizvandi M, Azadbakht A, Rostamnia S. Controlled uptake and release of imatinib from ultrasound nanoparticles $Cu_3(BTC)_2$ metal-organic framework in comparison with bulk structure. Journal of Colloid and Interface Science. 2016;**471**:112-117. DOI: 10.1016/j.jcis.2016.03.018

Section 4

Nanocomposite Materials: Sustainable Environment

Chapter 6

Nanocomposite Materials

Mousumi Sen

Abstract

Nanocomposites are the heterogeneous/hybrid materials that are produced by the mixtures of polymers with inorganic solids (clays to oxides) at the nanometric scale. Their structures are found to be more complicated than that of microcomposites. They are highly influenced by the structure, composition, interfacial interactions, and components of individual property. Most popularly, nanocomposites are prepared by the process within in situ growth and polymerization of biopolymer and inorganic matrix. With the rapid estimated demand of these striking potentially advanced materials, make them very much useful in various industries ranging from small scale to large to very large manufacturing units. With a great deal to mankind with environmental friendly, these offer advanced technologies in addition to the enhanced business opportunities to several industrial sectors like automobile, construction, electronics and electrical, food packaging, and technology transfer.

Keywords: nanocomposites, composites, phases, latex, disperse nanomaterials

1. Introduction

The nanoparticle includes the particles having size between 1 and 100 nm. These particles have different properties at their atomic level due to their size. This change in properties of nanoparticles is beneficial in many fields [1, 2]. Nanotechnology is one of the most interesting fields for researchers since the last century. Numbers of developments have been made since then in the field of nanotechnology. Nanoparticles can be classified as metal nanoparticles, non-metal ceramic nanoparticles, semiconductor nanoparticles, and a well-known type is carbon nanoparticles [3]. Nanoparticles have those chemical and physical properties which makes them very different from that of the corresponding bulk materials due to their small size and large surface to volume ratio. They attract much attention because of their potential applications in many fields including optics, electrics, magnetism, ceramics, and catalysis [4].

2. Composites

Composites are engineered or naturally occurring solid materials which results when two or more different constituent materials, each having its own significant characteristic (physical or chemical properties) are combined together to create a new substance with superior properties than original materials in a specific finished structure [5, 6]. They are commonly designed to offer wide range of properties and characteristics, some are as follows:

- Stiffness and strength
- Low coefficient of expansion
- Resistance against fatigue
- Ease in manufacturing complex shapes
- Simple repair of damaged structures
- Resistance to corrosion

3. Nanocomposites

Nanocomposites are those composites in which one phase has nanoscale morphology like nanoparticles, nanotubes, or lamellar nanostructure. They have multiphases, so are multiphasic materials, at least of the phases should have dimensions in the range of 10–100 nm. To overcome the limitation of different engineering materials now-a-days, nanocomposites are emerged to provide beneficial alternatives. Nanocomposites can be classified on the basis of their dispersed matrix and dispersed phase materials [7]. With the help of this rapidly expanding field, now-a-days, it has been possible to generate many exciting new materials with novel properties via innovative synthetic approaches. The properties of the so-called found not only depended on the properties of their originals, but also crucially on their interfacial and morphological characteristics. Of course, we cannot ignore the fact that sometimes it also happened that the newly generated property in the material is unknown to the parent constituent materials [8, 9]. Hence, the idea behind nanocomposite is to use building blocks with dimensions in nanometer range to design and create new materials with unprecedented flexibility and improvement in their physical properties.

4. Advantages of designing novel nanocomposites

Nanocomposites are the solid combination of a bulk matrix and nanodimensional phase(s) which differ in properties due to dissimilarities in structure and chemistry. Properties that have indicated substantial improvements:

- Mechanical properties (strength, bulk modules, withstands limit, etc.)
- Thermal stability
- Hinders flame and reduce smoke generations
- Permeability of gases, water, and solvents are reduced
- More surface appearance
- Improved electrical conductivity
- Increased chemical resistance
- Enhance optical clarity as compared to conventionally filled polymers

Among several nanocomposites, polymer-based nanomaterials are the most leading materials of current research and development. Characteristics like film forming ability, activated functionalities, and dimensional variability provide lots of benefit to polymer-based nanocomposites [10].

The potential risk of nanocomposites commonly occurs majorly in areas like

- Risk to health and environment
- Molecular manufacturing
- Societal risks
- Formation of nanocomposite materials:

Nanocomposites can be formed by blending inorganic nanoclusters, fullerenes, clays, metals, oxides, or semiconductors with numerous organic polymers or organic and organometallic compounds, biological molecules, enzymes, and sol-gel derived polymers (**Figure 1**).

Nanocomposite materials that are obtained by the combination of two or more separate building constituents in one material offers unique properties that plausibly arises from their small size, large surface area, and off course from the interfacial interaction between the phases. Their extra ordinary potential have been smoothly utilized to enhance the biological potential of many drugs, biomaterials, catalysts, and also in some high-value added materials [11].

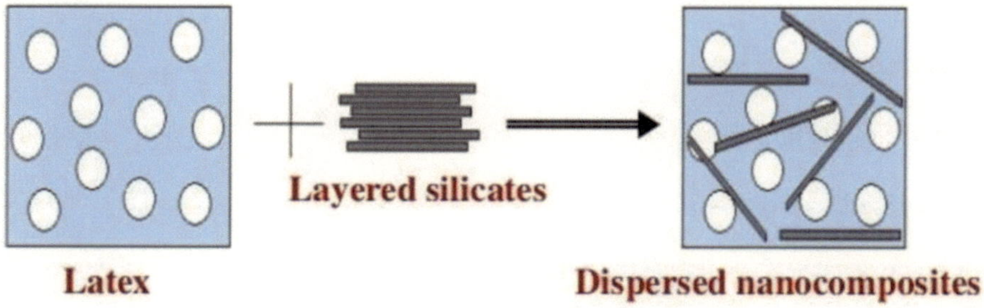

Figure 1.
Formation of nanocomposite materials.

5. Types of nanocomposites

Nanocomposite materials can be classified in the following way based on the presence or absence of polymeric material in the composite.

The nanocomposites in which the compositions do not contain any polymers or polymer-derived materials are called non-polymer-based nanocomposites (**Figure 2**). Non-polymer-based nanocomposites are also known as inorganic nanocomposites. They can be further classified into metal-based nanocomposites, ceramic-based nanocomposites, and ceramic-ceramic-based nanocomposites [12].

5.1 Non-polymer-based nanocomposites

5.1.1 Metal-based nanocomposites

Bimetallic nanoparticles being investigated in detail in the form of either of alloy or core-shell structures due to their improved catalytic properties and advancement

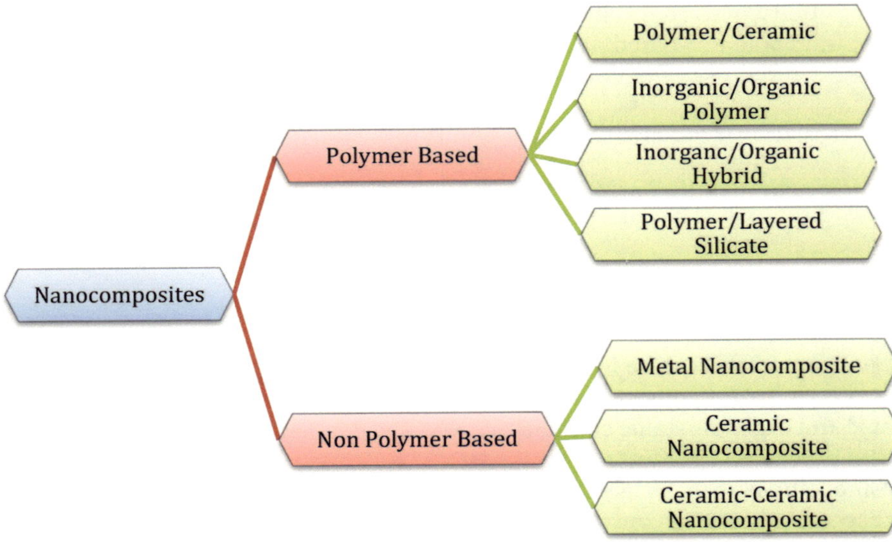

Figure 2.
Classification of polymer- and non-polymer-based nanocomposites.

in optical properties related to individual and differentiate metals [13]. They can be characterized by:

- Super plasticity,
- Lower melting points,
- Increased strength and hardness,
- Improved magnetic properties,
- Increased electrical resistivity, etc.

Non-polymer-based nanocomposites can be also classified as: metal/metal nanocomposites, for example Pt-Ru nanocomposites.

5.1.2 Ceramic-based nanocomposites

Ceramic-based nanocomposites are defined as ceramic composites with more than one solid phase, in which at least one of the phases has dimensions in the nanoscale range (<50–100 nm). In these types of composites, both the phases have combined magnetic, chemical, optical, and mechanical properties, for example hydroxyapatite/titania nanocomposites [14–16].

These can be characterized by:

- Better toughness
- Increased ductility
- Increased strength and hardness

5.1.3 Ceramic-ceramic-based nanocomposites

The non-polymer-based nanocomposites can be also classified as ceramic/ceramic nanocomposites which can be used in the area of artificial joint implants for fracture failures and it could promptly reduce the cost of surgery and would extend the mobility of the patient. The life spam would increase by 30 years, if the use of zirconia-toughened alumina nanocomposite implants is used effectively. The other example of ceramic/ceramic nanocomposites are calcium sulfate-biomimetic apatite nanocomposites [17].

The most promising prospects of both metal-based nanocomposites and ceramic-based nanocomposites are in the application of areas in dentistry in which the non-polymer-based nanocomposites or the inorganic materials that is metal or ceramics such as calcium phosphate, hydroxyapatite, and bioactive glass nanoparticles are very advantageous in alveolar bone regeneration and enamel substitution [18].

6. Polymer-based nanocomposites

The polymer or copolymer which contains nanoparticles or nanofillers dispersed in the polymer matrix is termed as poly nanocomposites. One dimension (1D) must be lying in the range of 1–50 nm and these possess several shapes like as platelets, fibers, spheroids, etc. Poly nanocomposites are in the category of multiphase systems such as, MPS namely composites, blends, and foams which can absorb about 95% of the production of plastics. So, these systems need controlled mixing, the achieved dispersion should be stable, dispersed phase should be oriented, and the compounding strategies which are involved for all MPS, which includes poly nanocomposites (PNC) is almost same [19].

Polymer nanocomposites are proposed as a class of materials with unique properties but, the most challenging property of PNCs is the complex interfacial areas in between the polymer matrices because of this small scale large specific area is created that highlight the importance of polymer-nanoparticle interactions. So, to achieve properties, such as, mechanical, thermal, optical, and electric, we need to analyze the intercalation process among the nanoparticles and polymer bases [20].

Polymer nanocomposites are known to be a class of reinforced polymer with a very low, i.e., less than about 5% of nanometric clay particles. These substances gained huge attention simultaneously from both the academic institution as well as from industrial sectors commonly in the area of nanocomposites. This is actually due to their drastically enhanced or improved thermal, mechanical as well as the barrier properties as compared to the micro- and also the conventional composites. These materials can be differentiated notably by: improved fire resistance and thermal stability, improved barrier properties, and increased recyclability [21].

However, despite of having so many advantages, it is still very much difficult to prepare a uniform dispersion between the filler and the matrix, as shown in **Figure 3**. Hence, unlikely, it reflects the lower mechanical as well as thermal properties in the produced nanocomposites.

6.1 Uses of polymer nanocomposites

Figure 4 shows the various uses of polymer nanocomposites irrespective of the nature of the field used. By the hydrolysis of tetraethyl-ortho-silicate, the hybrids made of poly rubber (dimethyl siloxane) and nanosilica can be given a specific shape like objects, such as golf balls (**Figure 5**). Many number of polymer nanocomposites for example, rubber, propylene, styrene butadiene rubber, and ethylene vinyl acetate are used in barrier applications.

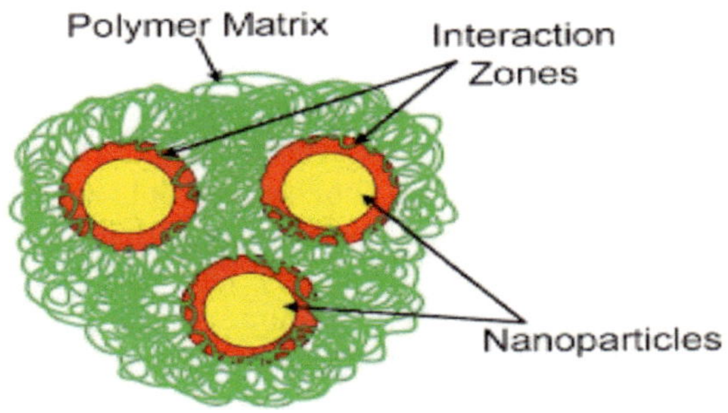

Figure 3.
Uniform dispersion between the filler and matrix in nanocomposites.

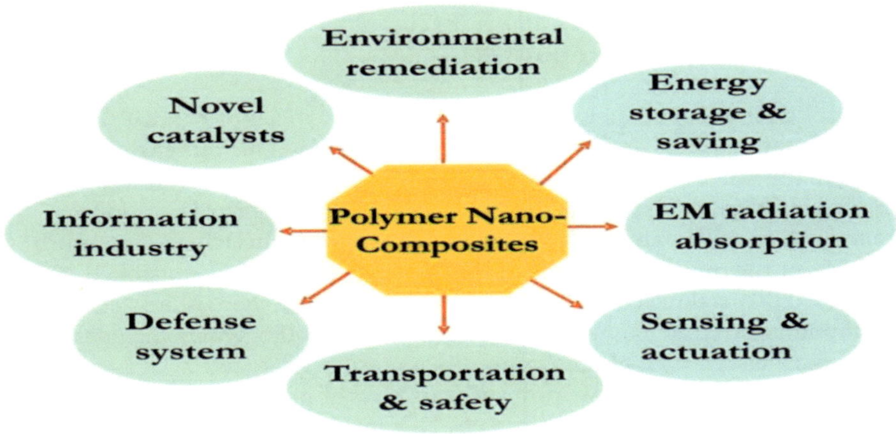

Figure 4.
Various uses of polymer nanocomposites.

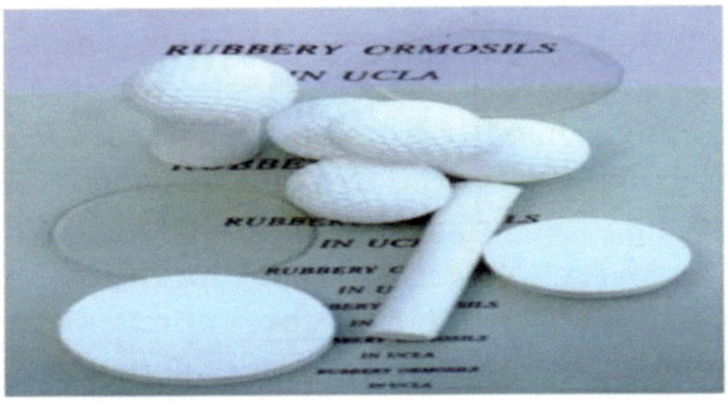

Figure 5.
Rubbery hybrids with different shapes.

They can act as a tremendous barrier for chemicals like toluene, sulfuric acid, and hydrochloric acid as well as for several gases such as, carbon dioxide, oxygen, and nitrogen [22, 23]. They are also utilized in chemical protective and surgical gloves as they have excellent solvent barrier properties in order for avoiding contamination from medicine.

Polymer nanocomposites are also used in food packaging, and the particular examples for food packaging includes processed cheese, meat, and dairy products also the medical containers for carrying blood collection tubes, baby pacifiers, and drinking water bottles. To enhance the barrier, mechanical properties and the life of the product clay-based polymer nanocomposites are been used in plastic bottles [24].

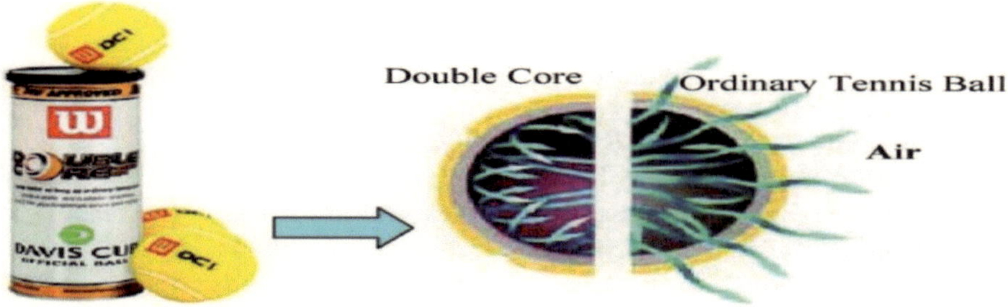

Figure 6.
The core of this Wilson tennis ball is covered by a polymer-clay nanocomposite coating.

Nanocomposites are also incorporated for beer bottle manufacture, so as to reduce many problems like biological and non-biological aspects, beer colloids instability, oxygen permeation, and change in taste due to light exposure. The double core Wilson tennis ball is the most recently commercialized sports goods (**Figure 6**), in which the coating of clay nanocomposites is done in order to maintain the internal pressure for a long period of time and also the core is coated by butyl rubber clay nanocomposite which doubled the shell life and acts a gas barrier.

6.2 Electronics and automobile sectors

In today's time, the biggest milestone is the incorporation or application of polymer/clay nanocomposites in the field of electronics and automobile sectors. Specifically to decrease the solvent transmission through polymers like elastomers, poly urethane, is the most impressive ability nanoclay incorporation. The poly nanocomposites help in the reduction of weight and processing cost so that they used by tire companies which are the major driving force for their usage. It is the naturally obtained materials which possesses low density. When the clay incorporated tires are compared with the ordinary ones then it is found that they have excellent mechanical properties and also improved gas barrier performance for tubeless tires uses [25]. Mostly for automobile tire manufacturing, styrene butadiene and natural rubber nanocomposites are preferred the most. It is due to their improved thermal properties and abrasion resistance that contribute to the long life of the tires. They have great applications in solar cells, transistors, battery manufacturing, etc.

6.3 Coatings

The most important modifying property of surfaces is coating. So, many methods and strategies are tried by the researchers to improve the surface properties of several products. The properties such as, excellent resistance for chemicals, better barrier properties, super hydrophobicity, and corrosion resistance are exhibited by nanoclay incorporated thermoset polymer nanocoatings [18]. The process parameters, such as dipping time, nature of surfactant, temperature, etc., determine the coating thickness. The thermoset polymer nanocoatings which are incorporated clay and nanosilver could improve the antibacterial properties and is used in medical sectors.

6.4 Bio-nanocomposites

The name "Bi-nanocomposites" is given as they are characterized as natural nanocomposite. To understand their essential roles in biological systems, their structures and properties are studied by biologists. Bio-nanocomposites are designed originally and are present to fulfill the needs of life and to meet surrounding environmental conditions so they can guarantee the living of the associated species. Natural materials are different in terms of structure and compositions but the design of bio-nanocomposites require biological molecules to consider them as synthetic building blocks, which is far more distant from the context of their own natural function. They are made of biopolymers and inorganic solids which has the dimension in the range of 1–100 nm. Due to their multidimensional properties such as antimicrobial activity, biocompatibility, and biodegradability they have several numbers of applications. The effective outcome of growing needs of bio-based polymers is the drastic reduction in the usage of fossil fuels. Bio-nanocomposites have easily replaced conventional non-biodegradable petroleum-based plastic as they are light weight and eco-friendly; they have become a sustainable that is future lasting material for use in high performance applications. As they are biocompatible, it makes them beneficial for biomedical applications and also makes them suitable for cosmetics and biotechnology applications. They have dominant significance in the future as green sustainable materials [26]. Bio-nanocomposites will act as substituents for the currently used petroleum-based polymers.

7. Uses of bio-nanocomposites

1. Bio-nanocomposites are used in cosmetics industries and also in the fabrication of implants, scaffolds, diagnostics, and biomedical devices.

2. They are also very beneficial as catalysts, contact lenses, and gas-separation membranes.

3. In the treatment of osteomyelitis, by the regeneration of tissue biopolymer-based nanocomposite is applicable.

4. Artificial bone implants involves nanostructured organic/inorganic nanocomposites which are useful in managing load-bearing bone grafts.

5. Using live cells of functionalized particles, controlled electrophoretic assembly of bioinorganic composite materials is done.

6. They are largely applicable in diagnostic, drug delivery, and tissue generation.

7. In industries, they are used as actuators. They are also used in the manufacturing of compostable bags as they are eco-friendly.

8. Applications of nanocomposites

Nanocomposites have been growing with a speedy rate so as their large number of applications. In the next 10 years, the worldwide production will exceed 600,000 tons in the following regions:

1. Superior strength fibers and films

2. UV protection gels

3. Drug delivery systems

4. New fire retardant materials

5. Anti-corrosion barrier coatings

6. Lubricant and stretch paints

Nanocomposites have also attracted the field of automotive and industrial applications by doing enhancements in especially the mechanical properties. They can be used or applied in the various vehicles types like engine covers, door covers, and timing belt covers. Other applications are usage as blades for vacuum cleaners, mower hood, covers for mobile phones, etc.

9. Future aspects

Modification of surface properties of nanoparticles by treating them with green agents for specific applications having specific improved microstructural properties like improved exfoliation, compatibility, and also thermal stability.

Advance optimization of the polymerization conditions during the preparation of the nanocomposites in order to get maximum output with minimal cost.

Detailed study on the effect of composition of the nanocomposites to build up the developed microstructures during the preparation activities.

Preparation of nanocomposites as well as their blends by using the materials like polymer blends along with the melt blending technologies. Hence, the advantages of the properties of the individual material as well as their coaction can be developed.

Using nanocomposites to make flexible batteries: "A nanocomposite of cellulous materials and nanotubes could be used to make a conductive paper. When this conductive paper is soaked in an electrolyte, a flexible battery is formed."

10. Conclusion

With the rapid development of nanotechnology in the past few years, the study of the nanocomposites has been increasingly become important in the development of new materials for advanced applications. To fulfill the growing needs of multifunctional materials, nanocomposites are the right choice as these are not only the versatile class of materials, but also have a high level of integrated association. It is a multidisciplinary field which includes the knowledge of scientific background as well as technological aspects to create macroscopic engineered materials obtained through nanolevel structures. These materials are suitable materials to meet the emerging demands arising from scientific and technologic advances. Outstanding potentials of nanocomposites can be exemplified by the massive investments from many companies and governments throughout the world. As a result, nanocomposites are expected to generate a great impact in world economy and business. The important aspects is that it provides plausible benefit to many of our industrial sector like electronics and electrical industry, chemical industry, transportation sectors, health care organizations, and above an all the protection of the environment. Hence, these are expected to have high impact on making the environment cleaner, greener, and safer in the coming years.

Acknowledgements

The author is very much thankful to the faculty members of AIAS, Amity University, Uttar Pradesh for providing necessary facilities and their constant encouragement to complete the above assessment.

Author details

Mousumi Sen
Amity Institute of Applied Sciences, Amity University, Noida, Uttar Pradesh, India

*Address all correspondence to: mosumi1976@gmail.com

References

[1] Yang C, Li W, Yang Z, Gu L, Yu Y. Nanoconfined antimony in sulfur and nitrogen co-doped three-dimensionally (3D) interconnected macroporous carbon for high performance sodium-ion batteries. Nano Energy. 2015;**18**:12-19

[2] Pan D, Wang S, Zhao B, Wu M, Zhang H, Wang Y, et al. Li storage properties of disordered graphene nanosheets. Chemistry of Materials. 2009;**21**:3136-3142

[3] Shao Y, Xiao J, Wang W, Engelhard M, Chen X, Nie Z, et al. Sodium ion insertion in hollow carbon nanowires for battery applications. Nano Letters. 2013;**13**:3909-3914

[4] Kaskhedikar NA, Maier J. Lithium storage in carbon nanostructures. Advanced Materials. 2010;**21**:2664-2680

[5] Shin WH, Jeong HM, Kim BG, Kang JK, Choi JW. Nitrogen-doped multiwall carbon nanotubes for lithium storage with extremely high capacity. Nano Letters. 2012;**12**:2283-2288

[6] Liu X, Antonietti M. Molten salt activation for synthesis of porous carbon nanostructures and carbon sheets. Carbon. 2014;**69**:460-466

[7] Wang J, Kaskel S. KOH activation of carbon-based materials for energy storage. Journal of Materials Chemistry. 2012;**22**:3710-23725

[8] Wang S, Xiao C, Xing Y, Xu H, Zhang S. Carbon nanofibers/nanosheets hybrid derived from cornstalks as a sustainable anode for Li-ion batteries. Journal of Materials Chemistry A. 2015;**3**:6742-6746

[9] Abdel Salam M, Mokhtar M, Basahel SN, Al Thabaiti SA, Obaid AY. Removal of chlorophenol from aqueous solution by multi-walled carbon nanotubes: Kinetic and thermodynamic studies. Journal of Alloys and Compounds. 2010;**500**:87-92

[10] Al-Johani H, Abdel Salam M. Kinetics and thermodynamic study of aniline adsorption by multi-walled carbon nanotubes from aqueous solution. Journal of Colloid and Interface Science. 2011;**360**:760-767

[11] Xin X, Wei Q, Yang J, Yan L, Feng R, Chen G, et al. Highly efficient removal of heavy metal ions by aminefunctionalized mesoporous Fe3O4 nanoparticles. Chemical Engineering Journal. 2012;**184**:132-140

[12] Khandoker N, Hawkins SC, Ibrahim R, Huynh CP, Deng F. Tensile strength of spinnable multiwall carbon nanotubes. Procedia Engineering. 2011;**10**:2572-2578

[13] Fam DWH, Palaniappan A, Tok AIY, Liedberg B, Moochhala SM. A review on technological aspects influencing commercialization of carbon nanotube sensors. 2011;**157**:1-7

[14] Gleiter H. Materials with ultrafine microstructures: Retrospectives and perspectives. Nanostructured Materials. 1992;**1**:1-19

[15] Braun T, Schubert A, Sindelys Z. Nanoscience and nanotechnology on the balance. Scientometrics. 1997;**38**:321-325

[16] Pandey JK, Kumar AP, Misra M, Mohanty AK, Drzal LT, Singh RP. Recent advances in biodegradable nanocomposites. Journal of Nanoscience and Nanotechnology. 2005

[17] Thostenson ET, Li C, Chou TW. Nanocomposites in context. Composites Science and Technology. 2005

[18] Vaia RA, Wagner HD. Framework for nanocomposites. Materials Today. 2004

[19] Fischer H. Polymer nanocomposites: From fundamental research to specific applications. Materials Science and Engineering. 2003

[20] Nalwa HS. Handbook of Nanostructured Materials and Technology. New York: Academic Press; 2000

[21] Ray SS, Bousmina M. Biodegradable polymers and their layered silicate nanocomposites: In greening the 21st century materials world. Progress in Materials Science. 2005;**50**:962

[22] Stankovich S, et al. Stable aqueous dispersions of graphitic nanoplatelets via the reduction of exfoliated graphite oxide in the presence of poly(sodium 4-styrenesulfonate). Journal of Materials Chemistry. 2006;**16**:155-158

[23] Qiuli Z, Zhenjiang J, Jun Z, Xicheng Z, Xinzhe L. Preparation of lanthanum oxide nanoparticles by chemical precipitation method. Materials Science Forum. 2012;**724**:233-236

[24] Basak Y, Kokuoz K, Serivalsatit BK, Olt G, McCormick E, John B. Er-doped Y_2O_3 nanoparticles: A comparison of different synthesis methods. Journal of the American Ceramic Society. 2009

[25] Mirosław Z, Kepinski L, Forget S, Chénais S. Preparation and Characterization of Lanthanum Oxide doped Barium Zirconate Titanate ($BaZr_{0.1}Ti_{0.9}O_3$; BZT) Ferroelectric Glass Ceramics. Springer Series in Optical Sciences. p. 175. DOI: 10.1007/978-3-642-36705-2_2

[26] Higgins TV. Improved optical properties in nanocrystalline Ce:YGG garnets via normal and reverse strike co-precipitation method. The Three Phases of Lasers: Solid-State, Gas, and Liquid. Laser Focus World. July 1995. pp. 73-85

Chapter 7

Novel Slow Release Nanocomposite Fertilizers

Muthuraman Yuvaraj and
Kizhaeral Sevathapandian Subramanian

Abstract

Nanotechnology deals with atom-by-atom manipulation and the strategies and products developed are quite precise. Despite the fact that the nanotechnology is noticeably exploited in the subject of energy, environment and health, the research is agricultural sciences had just scratched the surface. However, the potentials of nanotechnology in agricultural sciences had been reviewed. Among the applications, nanofertilizers technology is very revolutionary and known to exhibit economic advantage if the products advanced are economically feasible and socially sustainable. These nano fertilizers are pronounced to reduce nutrient loss due to leaching, emissions, and long-term incorporation by soil microorganisms.

Keywords: nanofertilizer, slow release, composite, use efficiency

1. Introduction

Today agriculture in the world is facing major tasks are reduction in yield, shrinking in the cultivable land due to globalization, less efficiency of nutrient, lack of nutrient availability and uptake is poor in soil, decreasing organic matter in soil, deficiency of water accessibility. In this critical situation it is more challenging to produce adequate food to feed the increasing populaces, which is projected to pass 9 billion by 2050. The nanofertilizer is ecologically safe and increase soil fertility, crop productivity and nutrient use efficacy. Nanofertilizer deals with atom-by-atom manipulation and the processes and products evolved are quite precise. Despite the fact that the nanotechnology is greatly exploited in the field of energy, environment and health, the research is agricultural sciences had just scratched the surface. Conversely, the importance and potentials of nanotechnology in agricultural sciences had been reviewed [1]. The nanofertilizers technology is very inventive and known to show economic benefit if the products evolved are economically viable and socially maintainable. These customized nanofertilizers are reported to decrease nutrient loss due to leaching, emissions in soil ecosystem [2].

Nano based encapsulated and slow release fertilizers increase the uptake of nutrients, enhance fertility of soil and decreasing toxic effects associated with over application of fertilizer. In Tamil Nadu Agricultural University, Coimbatore, various nano-zeolite based fertilizer research carried out with Nitrogen [3, 4], phosphorous [5], potassium [6], sulfur [7, 8] have been synthesized, characterized and examined in various crops.

Considering the above referred research, there is a crucial requirement to increase smart nanofertilizer can steadily release chemical substances to exact focused places and effectively control nutrient insufficiency. Nano based smart delivery system of nutrient to crop regulated slow release, target oriented and need based [9]. The important crop based nanofertilizer or nano formulation were produced which effectively increase growth and yield of the crops without create any harmful effect in environment ecosystem [10]. Nano fertilizer are less in size, more surface area, high sorption and desorption ability, slow release of nutrient for prolonged time. Conversely, the nanotechnology is new emerging science while using nanofertilizer to crop we have to conform safety measures to environment.

2. Technology of nano fertilizers

The nano-fertilizer denotes in nano scale range to deliver nutrients to plant and also present invention which substitutes conventional fertilizer the nanofertilizer release and uptake of nutrients in the soil and crop is high [10]. The nano fertilizer will improve absorption of nutrient, potentially enhance photosynthesis, enhances the crop production [11]. The encapsulation technique is used to hold nutrient inside the carrier with polymer and steadily release nutrient to crop. The zeolite based nano porous fertilizer utilization and interest will increasing within young researchers in nano technology field [12, 13] nanofertilizer can enable nutrient carriage to the rhizosphere region and minimize nutrient loss and further improve use efficiency of applied fertilizer.

The nano fertilizers work carried out by [14] reported that using silica nano mesoporous particle to encapsulate urea and produce nano nitrogen slow release fertilizer [15] found that apatite as a source of nano phosphatic fertilizer will reduce the hazard and eutrophication problem in water ecosystem. The nano size in nature of fertilizer will enter into the plant cell is very easy without creating any ill effect [16] research reported that chitosan biodegradable polymeric molecule has been used as a source to produce nitrogen, phosphorous and potassium based nanofertilizer.

2.1 Important benefits of nanofertilizers

Nanofertilizer innovative needed products for fertilizer industry and it's having higher surface area and auspicious picking for improving the quality and quantity of plants and seeds grown for consumption, to minimize production cost as well as ecofriendly to sustainable food production. The nanofertilizer are smaller in size, shape, charge of particle this will synthesized based on crop specific and demand oriented. Abundant particle like silver, titanium, zeolite, copper, silica, aluminum, carbon, zinc, and nitrogen based nanofertilizer is available.

Generally nanofertilizers are slow release: over a period of time the nutrient will available to crop at entire life cycle. Quick release: the outer most shell of the nano particle it breaks easily and quick release of nutrient in to the soil. Specific release: some specific chemicals molecules involve to break shell of the nano particle. Moisture release: in the presence of water molecules in nanofertilizer release nutrient in easy manner. Heat release: at a particular temperature nano particle get released. pH release: specific alkaline or acidic condition favor the slow release of nanofertilizer. The nano composite and fertilizers efficiently reduce nutrient loss from environment and increase use efficiency of nutrient [17, 18] found that nanofertilizer play important in agriculture production up to 35–40% to effectively reduce chronic problem, eutrophication, and nano fertilizer are alternative for conventional fertilizer.

2.2 Possessions of nanofertilizer

- **Surface area:** The nanofertilizers possess small particle size which causes increased surface area. Increased surface area raises the nanofertilizer reactivity with other compounds thereby increases the nutrient use efficiency and nutrient uptake.

- **Solubility:** Nano fertilizer with solvents as water possess increased solubility. Excessive solubility of nanofertilizers increases nutrient bioavailability in soil by solubilization and spreading of insoluble nutrient in soil.

- **Particles size:** Nanofertilizer has particle size of much <100 nm. Due to decreased particle size the nanofertilizers diffuses into the plants and increases the uptake of nutrients

- **Encapsulation of nanofertilizer:** Encapsulated nanofertilizer increases the available and nutrient uptake by plants [19]. Zeolite-based encapsulated nanofertilizers enhance availability of zinc and nitrogen to prevent from denitrification, volatilization, and leaching of nutrient in the soil.

- **Controlled release of fertilizers:** Controlled release of fertilizer decreases the toxicity of fertilizer. In peanut seeds Zinc oxide nano fertilizer produce increased growth of root and percentage of germination than bulk zinc sulphate due to its controlled release nature [20].

- **Nutrient uptake efficiency:** Nanofertilizers as increased uptake efficiency and limited leaching loss of applied fertilizers [21].

2.3 Nanofertilizer prepared based on zeolite

Zeolite mesoporous particle is potentially used for synthesis of nanofertilizer and its having higher surface area (900 m^{-2} g^{-1}) of zeolite, making zeolite an extraordinarily effective ion exchange [22]. The surface area, excessive nutrient absorption ability, water holding capacity and internal micro pore numbers is high in nano sized zeolite mineral due to having desirable physical and chemical properties. Zeolite acts as a carrier of nitrogen, phosphorous, potassium and micronutrient fertilizers enhance the productiveness of crops [23, 24]. Accordingly, nanofertilizers prepared based on zeolite are successful of deliver nutrients up to 50 days in case of traditional fertilizer like urea ended with 10–12 days [25, 26] pronounced that ammonium and potassium encapsulated with clinoptilolite it will increase the solubility of nutrient to the crop reported that nano clay like zeolite and montmorillonite carrying nitrogen are ability to deliver prolonged period of time (>1000 h) than conventional fertilizers (<500 h).

The increase nitrogen use efficiency by utilizing adsorbent of nano zeolites. The nitrogen use efficiency of conventional urea rarely exceeds 30–35% and nano zeolites has massive viable to normalize the discharge of nitrogen and nano zeolite encapsulated urea supports adsorption of nitrogen in higher zeolite mesoporous structure. The nitrogen content of zeourea and nano-zeourea confined 18.5 and 28% respectively release nitrogen 34–48 days in case of urea the nitrogen release arrest within 4 days [27, 28] proven that mixing urea with zeolite and sago waste water has extremely good advantage over urea alone as the combination increase the ammonium and available nitrate ions. The zeoponic is a plant grown with zeolite as a substrate and release demand driven nutrient delivery system [29]. The release of phosphorous from unmodified fertilizer loaded with zeolite and surface modified zeolite from strong potassium dihydrogen phosphate was once performed the use of the constant flow percolation reactor. The phosphorous supply from surface modified zeolite used to be available even after 1080 h of continuous percolation, while phosphorous from potassium dihydrogen phosphate was once exhausted within 264 h [30].

The nanoparticles, nano-zeolite as higher surface area due to this property it release fertilizers and anionic sulphate in slow and constant manner [31]. The pure ammonium sulphate and surface modified sulfur nano-zeolite were exposed to test nutrient release pattern by utilizing percolation reactors. The research data obviously designate that all of the available sulphate in pure ammonium sulphate is exhausted after 384 h while the launch of SO_4^{2-} from sulphate loaded surface modified nano-zeolite is sustained even after 912 h, with concentrations ranging from 47.56 to 8.27 $\mu g\ g^{-1}$. The surface modified nano sulfur is confirmed effective sulphure nanofertilizer as compared to conventional sulfur [32].

3. Formulation and preparation of nanofertilizer

3.1 Nitrogen

The urea treated with hydroxyapatite nanoparticles is attained by controlled adding of phosphoric acid into a suspension of Ca $(OH)_2$ and urea, monitored by fast drying using spray dryer. The research found that release of urea from the nanohybrids with a 1:6 hydroxyapatite to urea ratio released urea 12 times more gradually associated to pure urea. Additionally, the nanohybrid confined very nearly the same quantity of available nitrogen as pure urea [33].

3.2 Phosphorus

The encapsulated unmodified zeolite potassium dihydrogen phosphate release phosphorus from fertilizer and the percolation reactor used to test release pattern of surface modified zeolite from soil. The research found that the phosphorus source from fertilizer-loaded surface modified zeolite was accessible 1080 h of constant percolation, however phosphorus from potassium dihydrogen phosphate was arrest 264 h. This study confirmed that surface modified zeolite act as a potential nano fertilizer for phosphorus.

3.3 Potash

Li and Zhang [20] described that potassium encapsulated with zeolite as a controlled release fertilizer and observed the hot pepper growth parameter and potassium dynamics in soil. The high cation exchange capacity of the nano clays is produced when silica (Si^{4+}) is replaced by aluminum (Al^{3+}) increase negative charge in the clay lattice. This negative charge is composed by cations such as ammonium, sodium, calcium, and potassium, which are interchangeable with other cations. Potash fertilizer is directly involved in photosynthesis process, it assist stomata opening in leaves and water storage. Potash fertilizer are released slowly by using Polyacrylamide-based coated pellets.

The fertilizer contribute 35–40% of crop productivity along with seed and proper irrigation. The imbalance use of fertilization especially urea it may create surface water nitrate pollution and deficiency of nitrogen in soil. In the earlier few decades, use efficiencies of nitrogen, phosphorous and potassium fertilizers have continued stable as 30–35%, 18–20%, 35–40% respectively. To overcome multi-nutrient deficiencies, imbalanced fertilization, low fertilizer use efficiency and decreasing soil organic matter it is crucial to develop a nano-based fertilizer for smart delivery of nutrient to targeted site. The application of nanofertilizer in foliar spray of 640 mg/ha foliar application (40 ppm concentration) of nano phosphorous gave 18 kg/ha phosphorous equal yield of cluster bean and pearl millet in arid environment condition. The research data propose that stable fertilization can also be deliver through nanotechnological approach to meet out crop demand and fertilizers encapsulated in nanoparticles will enhance the uptake of nutrients [34].

4. Nutrient use efficiency and nanofertilizer

Enhancing nutrient use efficiency is a commendable goal and ultimate trial handled by agriculture fertilizer industry in worldwide. Presently nanofertilizer have involved with the experimental fields to increasing use efficiency of applied fertilizer. The nanofertilizer consist of higher surface area because lesser in size of the nanoparticle and have high reactivity, solubility in water. The nano encapsulation techniques considered as three ways: (a) nutrient can be encapsulation inside nanoporous particle, (b) A thin polymer can be used for outer coating (c) Can be released nanosize level fertilizer. Zeolite based nano encapsulated fertilizer is ability to release nutrient in slowly in to the crops and increase nutrient use efficiency [35].

In the conventional fertilizer the 50–70% of low in nitrogen use efficiency. New smart delivery systems of nano technological approach is enhance nitrogen availability and use efficiency. The fertilizer use effectivity in 10–25% for phosphorus. With nano-fertilizers rising as substitutes to traditional fertilizers, buildup of nutrient in soils and thereby eutrophication and drinking water impurity may additionally be eliminated. In fact, nanotechnology has opened up new opportunities to enhance nutrient use efficiency and limit charges of environmental protection. The encapsulation techniques such as manganese core shell will help to uptake and slow release of nutrient need based (5). Maximum number of agricultural soils in India have low native fertility and effective and continuous crop production on these soils needs regular nutrient efforts. The considerable available of nutrients for recycling through animal manures and crop residues is significantly insufficient to reimburse for the quantities uptake in crop production.

However, the use of conventional fertilizers in worldwide improved progressively over a period of time the use efficiency of nutrients applied as fertilizers continues to remain terribly low in phosphorous (15–20%) and micronutrients (2–5%) like zinc, iron, copper. When nutrient inputs are used incompetently then both cost of farming and threat of biosphere pollution rise. Thus, the economy and ecology highlights the obsessive need for more effective use of nutrients in crop production. Since, fertilizer nutrients are exclusive and used in huge quantities at national level, any rise in use efficiency will lead to a considerable cut in nutrient necessity and huge economic advantage at national level [36]. The slow-release properties of Zn to plants may be closely associated with higher yields. Nanotechnology has great potential in agriculture as it can enhance the quality of life through its application in fields like sustainable and quality agriculture, and improved and rich food for the community [37, 38].

5. Environmental and health situation of nanofertilizers

The utility of nanostructures or nanoparticles as agrochemicals (fertilizers or pesticides) is systematically being explored, before nanofertilizers may want to be used in agriculture or farming for a general farm practice. The homes of many nanoparticles are viewed to be of attainable hazard to human health, viz., size, shape, solubility, crystal phase, type of material, and exposure and dosage concentrations. However, specialist opinions indicate that food products containing nanoparticles available in the market are probably protected to eat, but this is an area that needs to be more actively investigated. To address the protection challenge element research are required to know the effect of nanoparticles within the human body once exposed through nanofood. Researchers have to assess and improve suited evaluation techniques to investigate the impact of nanoparticles and nanofertilizers on biotic and abiotic factors of ecosystem. Among the various issues, the accumulation of nanomaterials in environment, edible part of plants would possibly be the necessary issues earlier than use in agriculture.

6. Conclusion

World population is increasing geometrically its great agricultural challenge for feed the developing population with nutritious food. The biotic and abiotic constraints which limits the agricultural productivity furthermore has an effect on human health and use of exclusive nanofertilizers to improving crop production in agriculture. Consequently, it is required to attentively study the association of nanoparticle and crop microbiome. Supplementary, in order to recognize the interface of nanoparticle with soil and environment ecosystem. Investigational confirmation of the allowable use of nanofertilizer quantity within safety limits need to be described. The interface of nanomaterials with soil and plants varies with the type of nanofertilizer the applied attention of nanoparticle the time of treatment, plant genotype and the stage of growth. Regardless of these possible benefits, the recommendation of nanofertilizer in crop enhancement could come with hazards for the environment non-target plants, useful soil organism affected if nano-materials are misrepresented.

Author details

Muthuraman Yuvaraj[1]* and Kizhaeral Sevathapandian Subramanian[2]

1 Agricultural College and Research Institute, Tiruvannamalai, Tamil Nadu, India

2 Tamil Nadu Agricultural University, Coimbatore, India

*Address all correspondence to: yuvasoil@gmail.com

References

[1] Ahmed OH, Hussin A, Mohd Hanif Ahmad H, Boyie Jalloh M, Abd Rahim A, Muhamad Majid N. Ammonia volatilization and ammonium accumulation from urea mixed with zeolite and triple superphosphate. Acta Agriculturae Scandinavica, Section B. 2001;**58**(2):182-186

[2] Anderson K. Economic impacts of policies affecting crop biotechnology and trade. New Biotechnology. 2010;**27**:558-564. DOI: 10.1016/j.nbt.2010.05.012

[3] Andow D, Hutchison W. Bt-Corn Resistance Management. Now or Never: Serious New Plans to Save Natural Pest Control. Cambridge, MA: Union of Concerned Scientists; 1998. pp. 18-64

[4] Andrews RD, Shaw JW. 2010. Available from: http://www.zeoponix.com/new-page-5.htm

[5] Antoniou M. Genetically engineered food—Panacea or Pandora's box. Nutrition Today. 1996;**6**:8-11

[6] Apel A. The costly benefits of opposing agricultural biotechnology. New Biotechnology. 2010;**27**:635-675

[7] Bansiwal AK, Rayalu SS, Labhasetwar NK, Juwarkar AA, Devotta S. Surfactant-modified zeolite as a slow release fertilizer for phosphorus. Journal of Agricultural and Food Chemistry. 2006;**54**:4773-4779

[8] Bao-shan L, Shao-qi D, Chun-hui L, Li-jun F, Shu-chun Q, Min Y. Effect of TMS (nanostructured silicon dioxide) on growth of Changbai larch seedlings. Journal of Forest Research. 2004;**15**:138-140

[9] Cui HX, Sun CJ, Liu Q, Jiang J, Gu W. Applications of nanotechnology in agrochemical formulation, perspectives, challenges and strategies. In: International Conference on Nanoagri. Brazil: Sao Pedro; 2010. pp. 28-33

[10] DeRosa MC, Monreal C, Schnitzer M, Walsh R, Sultan Y. Nanotechnology in fertilizers. Nature Nanotechnology. 2010;**32**(5):1234-1237

[11] Fageria NK. Influence of micronutrients on dry matter yield and interaction with other nutrients in annual crops. Pesquisa Agropecuária Brasileira. 2002;**37**:1765-1772

[12] Ghafariyan M, Malakouti H, Dadpour MJ, Stroeve MR, Mahmoudi P. Effects of magnetite nanoparticles on soybean chlorophyll. Environmental Science and Technology. 2013;**47**:10645-10652

[13] He F, Zhao DY, Paul C. Field assessment of carboxymethyl cellulose stabilized iron nanoparticles for in situ destruction of chlorinated solvents in source zones. Water Research. 2010;**44**(7):2360-2370

[14] Hossain Z, Mustafa G, Sakata K, Komatsu S. Insights into the proteomic response of soybean towards Al_2O_3, ZnO, and Ag nanoparticles stress. Journal of Hazardous Materials. 2016;**304**:291-305

[15] Jaberzadeh A, Moaveni P, Moghadam HRT, Zahedi H. Influence of bulk and nanoparticles titanium foliar application on some agronomic traits, seed gluten and starch contents of wheat subjected to water deficit stress. Notulae Botanicae Horti Agrobotanici Cluj. 2013;**41**:201-207

[16] Jayvanth Kumar U, Vijay Bahadur S, Prasad VM, Shukla PK. Effect of different concentrations of Iron oxide and zinc oxide nanoparticles on growth and yield of strawberry (*Fragaria x ananassa* Duch) cv. Chandler. International Journal of

Current Microbiology and Applied Sciences. 2017;**6**(8):2440-2445

[17] Johnston ML. Soil Chemical Analysis. New Delhi: Prentice Hall of India Private Ltd.; 2010. pp. 56-70

[18] Lal R. Soils and India's food security. Journal of the Indian Society of Soil Science. 2008;**56**(2):129-138

[19] Latifah O, Ahmed OH, Nik Muhamad AM. Reducing ammonia loss from urea and improving soil exchangeable ammonium and available nitrate in non-waterlogged soils through mixing zeolite and sago (*Metroxylon sagu*) waste water. International Journal of Physical Sciences. 2011;**6**(4):866-870

[20] Li Z, Zhang Y. Use of surfactant modified zeolite to carry and slowly release sulfate. Desalination and Water Treatment. 2010;**21**:73-78

[21] Lin D, Xing B. Root uptake and phytotoxicity of ZnO nanoparticles. Environmental Science & Technology. 2008;**42**:5580-5585

[22] Liu R, Lal R. Synthetic apatite nanoparticles as a phosphorus fertilizer for soybean (*Glycine max*). Scientific Reports. 2014;**4**:5686

[23] Mahajan P, Shailesh K, Dhoke RK, Anand K. Effect of nanoparticles suspension on the growth of mung (Vigna radiata) seedlings by foliar spray method. Nanotechnology. 2013;**3**:4052-4081

[24] Mahmoodzadeh H, Nabavi M, Kashefi H. Effect of nanoscale titanium dioxide particles on the germination and growth of canola (*Brassica napus*). Journal of Ornamental and Horticultural Plants. 2013;**3**:25-32

[25] Malhi SS, Haderlin LK, Pauly DG, Johnson AM. Improving fertiliser use efficiency. Better Crops. 2002;**86**:22-25

[26] Manikandan A, Subramanian KS. Fabrication and characterisation of nanoporous zeolite based N fertilizer. African Journal of Agricultural Research. 2014;**9**(2):276-284

[27] Markovich A, Takac A, Illin Z, Ito T, Tognoni F. Enriched zeolites as substrate component in the production of paper and tomato seedling. Acta Horticulturae. 1995;**39**(6):321-328

[28] Meena DS. M.Sc. (Agri.) thesis, Dharwad, Karnataka (India): University of Agricultural Sciences; 2015

[29] Mishra V, Mishra RK, Dikshit A, Pandey AC. Interactions of nanoparticles with plants, an emerging prospective in the agriculture industry. In: Ahmad P, Rasool S, editors. Emerging Technologies and Management of Crop Stress Tolerance, Biological Techniques. Vol. 1. USA: Academic Press; 2014. pp. 159-180

[30] Mohanraj J. Effect of nano-zeolite on nitrogen dynamics and greenhouse gas emission in rice soil eco system [M.Tech. thesis]. Coimbatore: Tamil Nadu Agricultural University; 2013

[31] Mukhopadhyay D, Majumdar K, Patil R, Mandal MK. Response of rainfed rice to soil test-based nutrient application in Terai alluvial soils. Better Crops. 2008;**92**:13-15

[32] Naderi MR, Danesh Shahraki A. Nanofertilizers and their role in sustainable agriculture. International Journal of Agriculture and Crop Sciences. 2013;**5**:2229-2232

[33] Nair R, Varghese SH, Nair BG, Maekawa T, Yoshida Y, Kumar DS. Nanoparticulate material delivery to plants. Plant Science. 2010;**179**:154-163

[34] Pickering HW, Menzies NW, Hunter MN. Zeolite rock phosphate-a novel slow release phosphorus fertiliser

for potted plant production. Scientia Horticulturae. 2002;**9**(4):333-343

[35] Prasad TNV, Sudhakar KVP, Sreenivasulu Y, Latha P, Munaswamy V, Raja Reddy K, et al. Effect of nanoscale zinc oxide particles on the germination, growth and yield of peanut. Journal of Plant Nutrition. 2012;**356**:905-927

[36] Raliya R. Application of nanoparticles on plant system and associated rhizospheric rhizobacteria. Digest Journal of Nanomaterials and Biostructures. 2012;**4**:587-592

[37] Yuvaraj M, Subramanian KS. Controlled-release fertilizer of zinc encapsulated by a manganese hollow core shell. Soil Science and Plant Nutrition. 2015;**61**(2):319-326. DOI: 10.1080/00380768.2014.979327

[38] Yuvaraj M, Subramanian KS. Development of slow release Zn fertilizer using nano-zeolite as carrier. Journal of Plant Nutrition. 2018;**41**(3):311-320. DOI: 10.1080/01904167.2017.1381729

Chapter 8

Graphene Oxide-Based Nanohybrids as Pesticide Biosensors: Latest Developments

Navin Kumar Mogha

Abstract

Graphene is the most significant two-dimensional nanomaterial with sp^2 hybridized carbon atoms in a honeycomb arrangement with an extremely high surface area, excellent electrical properties, high mechanical strength, and advantageous optical properties and is relatively easy to functionalize and mass produce. Various inorganic nanoparticles incorporated with graphene, such as gold, silver, and palladium nanoparticles are brought into sharp focus due to their catalytic, optical, electronic, and quantized charging/discharging properties. Graphene oxide-based nanohybrids are particularly well suited for biosensing applications and catalysis. Consequently, this area of research has grown to represent one of the largest classes within the scope of materials science and is rapidly becoming a key area in nanoscience and nanotechnology offering significant potential in the development of advanced materials in multiple and diverse applications. Here in this present chapter, synthesis, characterization of graphene oxide, and their nanohybrids are discussed thoroughly with their application in the field of pesticide biosensors. This chapter will help in a further understanding of graphene-based nanohybrids as a biosensing platform for their future applications in a sustainable environment.

Keywords: graphene oxide, pesticides, biosensors, nanohybrids, nanoparticles, chemical reduction

1. Introduction

The prevalence of harmful and toxic chemical compounds in the environment has become a serious issue in recent decades [1]. Contamination of foodstuffs, drinking water, and air with hazardous pollutants and other foreign substances are real and a direct threat to human health, whereas the accumulation of such contaminants in the human body and environment may lead to long-lasting, severe, and harmful effects after primary exposure [2]. Chemicals such as pesticides, plastic, lead, methylmercury, polychlorinated biphenyls, arsenic, toluene, rubber, and paper [3] play a key role in the economic growth of countries to fulfill their development objectives [4]. The term "pesticide" is defined as any chemical entity, which has the ability to kill the various kinds of pests including rodents, insects, fungi, weeds, etc. and henceforth categorized accordingly as rodenticides, insecticides, fungicides, and herbicides [5]. However, based on chemical composition, pesticides can be classified into five main groups as organochlorines, organophosphorus (OP), carbamates, pyrethrin, and pyrethroids compound. The unnecessary consumption of those agrochemicals has undesirable effects on the ecosystem, including a decreased population of beneficial insects as well as risks to vulnerable species and bird habitats. Pesticide pollution is becoming one of the most severe challenges of common public health around the globe because of their particular application in the agriculture sector to assure crop yield and productivity [6]. In some cases, acute poisoning may occur because of inappropriate handling that ultimately causes adverse health effects because of long-term and low-level exposures. The widespread diffusion of such toxic chemicals adversely affects a great part of the population. A large number of people, categorized by different patterns, ages, and degrees of exposure, are at increased risk to the adverse effects of these chemicals. Workers who are involved in the manufacturing and application of pesticides are at a considerable risk of exposure, which typically occurs among specific users in public health. In the agricultural sector, farmers may get direct exposure to pesticides during spraying across the agricultural fields [7, 8]. In the general population, individuals may be at a risk of pesticide exposure on a daily basis in food and drinking water or to pesticide drift in domestic areas adjacent to spraying areas [9]. Given their hazardous effect on human health and the environment, the prime concern should be of their rapid and reliable detection by a convenient method. Although various laboratory-based analytical methods such as colorimetry, capillary electrophoresis (CE), thin-layer chromatography (TLC), gas-liquid chromatography(GLC), high-performance liquid chromatography (HPLC), nuclear magnetic resonance (NMR) spectroscopy, mass spectrometry (MS), and enzyme-linked immunosorbent assays (ELISA) have been employed so far, but these suffer from one and the other drawback such as the use of expensive instrumentation, time-consuming process, and requirement of trained personnel [5]. Therefore, there is a dire need to develop sensitive, rapid, economically feasible, and easy-to-use methods for the detection of these compounds in the environment. Such efficient detection methods could be developed using biosensors that are used in a variety of applications for prompt and accurate detection of different analytes such as biomolecules and chemical compounds [10]. Various nanomaterials are generally categorized into nanoparticles, nanotubes, and nanocomposites, which can be generally employed for the diagnosis, degradation, and adsorption of chemical pesticides. Carbon nanomaterials or nanoparticles (NPs) have specific characteristics, including a high surface-to-volume ratio, good electrical conductivity, catalytic action, and beneficial biocompatibility and can be simply modified with functional groups, which has made them be often used in pesticide biosensors to boost analytical efficiency [11].

It is well known that graphite and diamond are its most common allotropic form of carbon found in nature. Graphite, which is found as a natural mineral, consists of sp^2 hybridized carbon atomic layers that are stacked collectively through weak attraction forces such as van der Waals forces. Single-layer out of these carbon atomic layers are packed in a two-dimensional honeycomb structure called as "graphene" termed coined by Boehm et al. [12]. It remains almost impossible to isolate graphene monolayer for several decades before Geim and Novoselov [13] who reported a scotch tape method or micromechanical cleavage method for the isolation of graphene monolayer from silicon oxide substrate. Discovery of graphene monolayer awarded Geim and Novoselov the Nobel Prize in Physics "for groundbreaking experiments regarding the two-dimensional material graphene".

Graphene, which consists of a one-atom-thick planar sheet comprising an sp^2-bonded carbon structure with exceptionally high crystal and electronic quality, is a novel material that has emerged as a rapidly rising star in the field of material science [14, 15].

Research-based on graphene oxide (GO) and graphene is an established interdisciplinary field associated with different disciplines such as physics, chemistry, material sciences, and nanotechnology with still a lot of emerging ideas to be developed. The result of working experience on other carbon allotropes leads to rapid discoveries of exceptional electronic, optical, and mechanical properties of graphene. In particular, its extraordinary charge carrier mobilities, thermal, and electrical conductivity, collective with high transparency and mechanical strength make graphene a suitable material for biosensing applications.

These exceptional physicochemical properties indicate its potential for delivering new tactics and critical developments in electrochemical sciences. For instance, a large number of analytic molecules can be attached to the large surface of electrically conductive graphene sheets leading to the development of the highly sensitive miniaturized device. Direct electron transfer between graphene and redox species creates new prospects for sensing applications. Consequently, graphene has lately fascinated the attention of the scientific community worldwide.

1.1 Graphene and graphene oxide

GO is considered as a precursor for obtaining graphene via chemical or thermal reduction methods. It consists of single-layer graphite oxide, having various oxygen-containing groups, whose structure has been proposed through several models over the years [16–20]. Oxygen functional groups have been identified as typically in the form of hydroxyl, epoxy groups and carboxy, carbonyl, phenol, lactone, quinone on the basal plane, and at the sheet edges, respectively [21–23]. However, due to ambiguity pertaining to the nature and distribution of the oxygen-containing functional groups (**Figure 1**) [24, 25], its nonstoichiometric atomic composition, and the absence of adequately sensitive analytical techniques for GO characterization, its precise structure cannot be fully elucidated. The difference between GO and pristine graphene is as a result oxygenated groups present in GO which affect its electronic, mechanical, and electrochemical properties. Hence, they account for the differences between GO and pristine graphene [26]. The covalent oxygenated functional groups in GO give rise to remarkable structure defects, which are associated

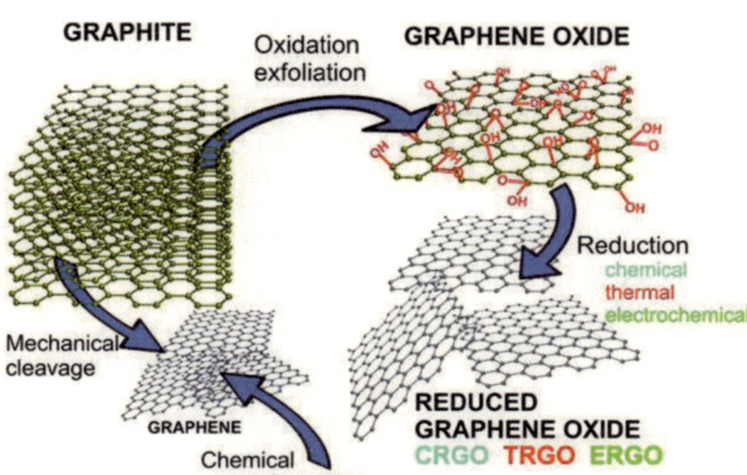

Figure 1.
A schematic illustration of methods for the preparation of graphene, GO, and rGO by means of mechanical cleavage, exfoliation, CVD, and reduction methods including chemical, thermal, and electrochemical methods from graphite. Reprinted with permission from Ref. [25], Published by Elsevier.

with some loss in its electrical conductivity [27], limiting the direct application of GO in electrically active materials and devices. In contrast, these functional groups can also be proved advantageous for exploiting GO in numerous other applications. The presence of polar oxygen-containing moieties gives GO a hydrophilic character making it dispersible in many solvents particularly in water [24, 28, 29]. Subsequent stable GO suspension can be used for preparing thin conducting films using spin coating, drop-casting, or spraying methods [23] for further to be used as electrodes.

Furthermore, well-known chemistry strategies can be used for the functionalization of GO using oxygen-containing groups as a site for chemical modification, which subsequently can be exploited for immobilization of various electroactive species via covalent or noncovalent bonds different application in sensing or catalysis. Thus, the physicochemical properties of GO can be tuned very easily by engineering its chemical composition [21, 30, 31].

Hydroxyl, epoxy, and carboxyl groups present in GO are covalently bonded to the carbon atom with sp^3 hybridization are termed as oxidized region, disrupting the extended sp^2 conjugated network of honeycomb lattice in graphene, which can be viewed as an unoxidized region [32, 33]. sp^3 hybridized carbon clusters with oxygen-containing groups are uniformly but randomly distributed either above or below the GO plane [34]. Various microscopic and spectroscopic techniques have been employed for an in-depth analysis of the structure of GO. For instance, atomic force microscopy (AFM) provides the apparent thickness of the single-layer GO sheet beside the number of layers present [33, 35–37].

In contrast, conductive AFM demonstrates electrical defects found in GO [38]. Lately, one of the significant breakthroughs in determining the structure of GO, high-resolution transmission electron microscopy (HRTEM) has been employed for direct imaging of lattice atoms and topological defects present in single layer of GO [39–41]. Erickson et al. [39] identified specific atomic scale topographies of the GO monolayers, consisting of three major portions *viz.* holes, graphitic regions, and high-contrast disordered regions having approximate area percentages of 2, 16, and 82%, respectively.

According to the author, excessive oxidation and sheet exfoliation lead to the release of CO and CO_2 consequently forming holes in GO. They also proposed that graphitic regions are a result of incomplete oxidation of basal plane having the preserved honeycomb structure of graphene, whereas the disordered region is rich in oxygen functionalities, such as hydroxyl, epoxides, and carbonyls with no order between them.

The chemical composition of GO and its oxygenated functionalities have been recognized through various spectroscopic techniques, which include solid-state nuclear magnetic resonance (SSNMR) [42–44], X-ray absorption near-edge spectroscopy (XANES) [45–49], Raman spectroscopy [45–49], X-ray photoelectron spectroscopy (XPS) [49] and Fourier transform infrared spectroscopy (FT-IR) [47, 50, 51]. Three main peaks around 60, 70, and 130 ppm are assigned to carbon atoms bonding to the epoxy group, hydroxyl group, and graphitic sp^2 carbon, respectively [44], can be seen in a typical solid-state ^{13}C magic-angle spinning NMR spectra of GO. Furthermore, three small additional peaks were also found at about 101, 167, and 191 ppm tentatively attributed to lactol, the ester carbonyl, and the ketone groups, correspondingly. XANES is another powerful tool for GO characterization, which provides information related to the degree of bond hybridization in mixed sp^2/sp^3-bonded carbon, the specific bonding arrangements of functional atoms, and graphitic crystal structure's degree of alignment inside GO [49].

Besides, Raman and FTIR spectroscopy data support the presence of oxygenated species in GO and its degree of oxidation. Raman spectrum of a GO displays two characteristic bands namely a D-band at ~1340 cm^{-1} and G-band at ~1580 cm^{-1} [52].

The G-band is a distinctive peak of all sp^2-hybridized carbon networks and due to first-order scattering from the doubly degenerate E_2g phonon modes of graphite whereas the D peak originates from structural imperfections and disorders produced by the addition of oxygenated groups on the carbon basal plane [52–54]. Hence, the intensity ratio of the D- and G-bands (I_d/I_g) points to the oxidation degree, disorders, and the size of sp^2 ring clusters in a matrix of sp^3- and sp^2-bonded carbon [53]. A significant decrease in I_d/I_g ratio was observed after thermal reduction, indicating a considerable regaining of conjugation in the graphitic structure after the defunctionalization of GO [45]. Functional groups can be recognized by the use of FT-IR spectroscopy and in the case of GO, it has reinforced the presence of hydroxyl (broad peak at 3050–3800 cm^{-1}), carbonyl (1750–1850 cm^{-1}), carboxyl (1650–1750 cm^{-1}), C=C (1500–1600 cm^{-1}), and ether or epoxide (1000–1280 cm^{-1}) groups [43, 47, 50].

Specific 2D structure and the presence of oxygenated functionalities are responsible for excellent properties of GO, which include electronic, optical, thermal, mechanical, and electrochemical properties along with chemical reactivity. Electronic properties like conductivity of GO sheets are dependent on its chemical and atomic structure; in particular, the degree of oxidation arises from disorders due to substantial sp^3 carbon. Generally, pristine GO films are insulating in nature with an energy gap in electron density of states, [55] as well as sheet resistance (Rs) about 10^{12} Ω $sq.^{-1}$ or higher [56]. This inherent insulating nature of GO is strongly associated with the amount of sp^3 C–O bonding, which acts as transport barriers, leading to the lack or interruption of penetrating pathways among the sp^2 carbon clusters. However, reduction of GO, whether chemical or thermal assists the transport of carriers, [57] helps to bring Rs down to several orders of magnitude and transforming the material into a semiconductor or finally into graphene-like material [58–60]. Reduced GO has conductivity up to ~1000 S/m, [61] and activation energy as 32 ± 5 kcal/mol, estimated by the use of resistivity and temperature-programmed desorption (TPD) measurements [62]. Additionally, GO exhibits unique optical properties photoluminescence (PL) [63] occurring near-UV-to-blue visible (vis) to near-infrared (IR) wavelength range. Applications of this property have been sought in biosensing, fluorescence tags, and optoelectronic applications [64, 65].

GO also demonstrates excellent electrocatalytic properties [66–68], such as the electrocatalytic activity of GO toward oxygen reduction and certain biomolecules [66], oxidation of hydrazine by reduced GO [67]. In addition to this, GO is capable of showing high electrochemical capacitance for application in ultracapacitors [68, 69]. As compared to carbon nanotubes, reduced GO exhibit higher electrochemical capacitance and cycling durability, wherever specific capacitance for reduced GO and carbon nanotubes was found to be ~165 and ~86 F/g, respectively [68].

The chemical reactivity of GO can be attributed to the presence of oxygenated functionalities and its disordered structure with defects. The reduction is the most important chemical reaction of GO, and it has been reduced by employing various approaches such as hydrazine, [70] sodium borohydride, [71] or hydroquinone, [72] in the liquid phase and the vapor phase using hydrazine/hydrogen [33, 52] or just by thermal annealing [52] or by using electrochemical techniques [73]. Chemical functionalization is another important chemical reaction involving GO, which includes the addition of other chemical groups to GO employing different chemical reactions. Oxygenated functionalities over GO surface play a very important role in its chemical reactions. Hence, it becomes an ideal approach to selectively chemical functionalize GO by utilizing reactions on these functionalities. Typically, covalent functionalization of GO can be realized using small molecules and polymers via activation, amidation, or esterification of either hydroxyls or carboxyl groups through coupling reactions [74–76]. For example, GO was made soluble in organic solvents by rendering a coupling reaction with octadecylamine via amide formation, where carboxyl functionalities of GO were first activated by $SOCl_2$ [74]. Ring-opening reactions can be used to functionalize epoxy groups by nucleophilic attack at α-carbon by the amine [77]. For example, octadecylamine attachment to GO surface [57], attachment of an amine group-containing ionic liquid through ring-opening reaction with epoxy groups on GO [77] making chemically functionalized GO more soluble in water as well as other organic solvents.

Noncovalent functionalization of GO is also known in addition to covalent modifications. Noncovalent modification of GO can be accomplished by various forces and interactions including hydrogen bonding, van der Walls interaction, π-π stacking, cation-π interaction [78, 79]. Doxorubicin hydrochloride (Dox)/GO hybrid was synthesized through noncovalent interactions using π-π stacking and hydrophobic interactions between the sp^2 carbon matrix and quinone functionality of Dox as primary noncovalent interactions. Additionally, strong hydrogen bonding between hydroxyl and amine groups of Dox with hydroxyl and carboxyl groups of GO also helps in covalent modification [78].

The usefulness of GO can be estimated from this fact that along with its applications in electronics and displays, it can also act as a carbocatalyst for assisting hydration and oxidation reactions [80–82]. GO can be used as a catalyst for oxidation of alcohols and alkenes besides hydration of alkynes into aldehydes and ketones [80]. Furthermore, GO has a broad range of oxidation reaction, for example, it can oxidize olefins to diones, methylbenzene to aldehydes as well as other dehydrogenations [83].

2. Graphene oxide-based nanohybrids

GO and reduced GO (rGO) themselves have many advantageous properties, but a substantial amount of work is being done to utilize these materials in combination with other nanomaterials such as nanoparticles or polymers. Based on their morphologies, graphene oxide/nanoparticle nanohybrids can be roughly divided into two classes: first where nanoparticles are grown or decorated upon sheets of GO and second, nanoparticles are wrapped in GO sheets. Particularly in the first type, graphene/nanoparticle nanohybrid can be synthesized by combining GO or rGO with different nanoparticles such as metal nanoparticles, metal oxide nanoparticles, quantum dots, or silica nanoparticles depending upon the application desired. A unique combination of the nanoparticles and GO/rGO makes a novel synergistic nanomaterial with enhanced and diverse properties. For example, decorated metal or metal oxide nanoparticles over GO surface modify the local electronic structure and hence the charge transfer behavior of graphene [84] resulting in improved catalytic behavior of this nanocomposite. Alternatively, enhancement in sensitivity and selectivity has been observed in sensors derived from the combination of graphene material and nanoparticles having good conductivity and catalytic behavior [85, 86].

Similarly, in graphene oxide/polymer nanohybrids, surface functionalities present on GO surface groups can assist the combination of GO with polymers or synthesis of the polymer by different polymerization techniques [87, 88]. A typical modification strategy includes covalent bonding, that is, “Grafting to” and “Grafting from” approaches, whereas noncovalent modification includes π-π stacking, electrostatic interaction, and hydrogen bonding [89–91]. Similarly, fabrication strategies of graphene oxide/polymer nanohybrids synthesis include in situ polymerization, melt compounding, latex blending, solution mixing, and electro polymerization [92, 93].

3. Synthetic methodologies for graphene oxide/nanoparticle nanohybrids

Graphene oxide/nanoparticle nanohybrids in which GO/rGO sheets are decorated with nanoparticles having dimensions ranging from few nanometers to a couple of hundred nanometers [94] can be attained by attaching different types of nanoparticles to the surface of GO sheets either by in situ method or by ex situ method. In situ method comprises growing nanoparticles on the surface of GO; however, in the case of ex situ method, pre-synthesized nanoparticles are immobilized over the surface of GO. The presence of defects and oxygenated functionalities makes GO an encouraging templates for the attachment, nucleation, and growth numerous metal (e.g., Au [87, 95], Ag [96], Pt [97], etc.) and metal oxide nanoparticles (e.g., Fe_3O_4 [98], TiO_2 [99], ZnO [100], SnO_2 [101], Cu_2O [102, 103], MnO_2 [104], NiO [105, 106], La_2O_3, [107, 108], etc.). Subsequent graphene oxide/nanoparticle nanohybrid offers several unique and beneficial properties for various applications depending on individual characteristics showed by nanoparticles immobilized upon GO.

The following section includes the different methods for the preparation of graphene oxide/nanoparticles nanocomposites, for example, chemical reduction, hydrothermal route, and electrochemical method or ex situ synthesis, while primarily focusing on individual characteristics and advantages of each technique correlated to the properties of resulting graphene oxide/nanoparticle nanohybrids.

3.1 Chemical reduction method

Graphene oxide/metal nanoparticle nanohybrids are mostly synthesized by chemical reduction of their metal salt precursors such as $HAuCl_4$, $AgNO_3$, and K_2PtCl_4 utilizing reducing agents such as sodium citrate, ethylene glycol or polyethylene glycol, and sodium borohydride [96, 109], positively charged metallic salts get nucleated on negatively charged functional groups of GO which results in the growth of metal nanoparticles on its surface, while reducing GO to rGO, preserving the excellent electrical properties of rGO. Moreover, the density of metal nanoparticles can also be controlled by tuning the density of oxygenated functionalities on GO.

Chemical reduction technique is the most basic method for the preparation of Graphene oxide/noble metal nanoparticle nanohybrids. In particular, gold (AuNP) and silver nanoparticles (AgNPs) are among the most comprehensively studied nanomaterials with a wide range of biomedical applications such as diagnostics, imaging, drug delivery [110]. High biocompatibility and surface plasmon resonance are some of the very unique properties of noble nanoparticles making them of particular interest. These properties can be tuned to desired values according to the shape and size of the nanoparticles [111]. Furthermore, graphene oxide/noble metal nanoparticle nanohybrids are able to show SERS in addition to enhanced catalytic activity [112]. Reduced graphene oxide/AuNPs are the most common and utilized nanocomposites, which can be prepared by mixing $HAuCl_4$ with GO and sodium citrate, followed by reduction using $NaBH_4$ to form AuNPs while reducing GO to rGO [113, 114]. Similarly, instead of using $HAuCl_4$, $AgNO_3$ is used for reduced graphene oxide/AgNPs composite synthesis [112, 114]. In a similar way, reduced graphene oxide/platinum nanoparticle or reduced graphene oxide/palladium nanoparticle nanohybrids are formed by mixing graphene oxide with chloroplatinic acid (H_2PtCl_6) or tetrachloropalladic acid (H_2PdCl_4), followed by reduction with ethylene glycol or any other reducing agent.

3.2 Hydrothermal methods

Another very common method for synthesizing inorganic nanoparticles is the hydrothermal method. This method gives nanoparticles with high crystallinity and narrow size distribution over graphene oxide. Moreover, there is no need for postannealing or calcination for reduced graphene oxide/metal nanoparticle nanohybrids. In general, the growth of nanocrystals is induced by high temperature and pressure, which is also responsible for the conversion of GO to rGO during the process. However, in most cases reducing agents are also added to make sure a complete reduction of GO [115].

The most common nanohybrids synthesized by the hydrothermal method are reduced graphene oxide/ metal oxide nanoparticle nanohybrids which include ZnO [116], TiO_2 [117], Fe_3O_4 [118], SnO_2 [119], etc.

Reduced graphene oxide/metal oxide nanoparticle hybrids illustrate their specific properties such as higher capacitance, which depends upon nanoparticle size, shape, and crystallinity; also, it helps in the suppression of restacking and agglomeration in graphene oxide sheets. Furthermore, these nanocomposites also exhibit enhancement in electron conductivity, high surface area as compared to GO or graphene, also shortened route for ion transfer, which in all responsible for their higher electrochemical activity. For instance, reduced graphene oxide/SnO_2 nanosphere nanohybrid exhibited significantly enhanced formaldehyde sensing performance compared to the pristine SnO_2 nanospheres [119]. Alternatively, reduced graphene oxide/magnetic nanoparticle nanohybrid has been prepared using $FeCl_3$ as an iron source and ethylene glycol as a reducing agent [120]. Resulting nanohybrid displayed outstanding electrical conductivity as well as magnetic properties. Similarly, chalcogenide quantum dots, for example, CdS [121], ZnS [122], Cu_2S [123], and MoS_2 [124], etc. have been successfully immobilized on graphene oxide exploiting hydrothermal methods.

3.3 Electrochemical deposition method

The electrochemical deposition method is a very simple, low cost, fast, easy to miniaturize, highly stable, reproducible, and green technique for preparation of graphene oxide/nanoparticle composite [125]. The advantage of this technique is that the size and shape of the nanoparticles to be deposited can be precisely controlled using varying the conditions of electrochemical deposition. Electrochemical deposition methods have been established for the fabrication of a vast variety of graphene oxide/ metal nanoparticle composites for noble metals like Au, [126] using cyclic voltammetry (CV), which helped in fabricating an electrode for the determination of trace amount As(III) employing square wave anodic stripping voltammetry, Ag, [127] for carrying out the oxidation of different amino acids such as glycine, alanine, leucine, aspartic and glutamic acids using cyclic voltammetry and amperometric techniques. Similarly for Pt [128], Pt nanoparticles embedded rGO on glassy carbon electrode are utilized to carry out electrooxidation of formic acid. Generally, a typical electrochemical deposition experiment is consisting of three basic steps, that is, assembly of graphene oxide sheets on the electrode, graphene oxide-coated electrode immersion in an electrolytic solution of selected metal precursors, and potential applied across electrodes. A majority of research has concentrated on using electrochemical deposition methods for synthesizing graphene oxide/metal nanoparticle composite, but there are some reports for preparation of graphene oxide/metal oxide nanoparticle composite synthesis by the same technique. For instance, Cl-doped n-type Cu_2O nanoparticles with a direct band gap of ca. 2.0 eV [128] have been deposited on rGO electrodes with a subsequent carrier concentration of up to $1 \times 1020\ cm^{-3}$ [129].

3.4 Graphene oxide/encapsulated nanoparticles

Flexible and 2D sheet-like structure of graphene oxide and its derivatives help in wrapping or encapsulating nanoparticles in the range from 100 nm to few micrometers.

GO and rGO sheets are most commonly used for nanoparticle encapsulation due to their hydrophilic nature and ease of fabrication. Noncovalent bonds are responsible for this type of encapsulation; for instance, modification of nanoparticle surface with a positive charge is used for electrostatic interaction with negatively charged GO [130, 131]. Encapsulation of a variety of nanomaterials, for example, polymer, inorganic nanoparticles, metal, and metal oxide nanoparticles, can be achieved by controlling the cracked size of GO and rGO, thus obtained composite offer enhanced properties and additional advantages. For example, enhancement in electrical, optical, and electrochemical properties has been observed for graphene oxide encapsulated nanoparticles, also suppression of aggregation of small nanoparticles [132, 133]. Moreover, leaching of nanoparticles is reduced in graphene oxide encapsulated nanoparticles due to the high amount of contact between GO and nanoparticles, making them more stable. Several reports have revealed the encapsulation of metal oxide nanoparticles with graphene oxide. For example, rGO encapsulated cobalt oxide nanoparticles have shown a very high reversible capacity (1000 mAh g^{-1}) over 130 cycles, much more than the normal cobalt oxide nanoparticles used for capacitors [131]. Moreover, rGO encapsulated $Co3O_4$ nanofibers-based sensor exhibited an excellent sensitivity with a fast response and recovery to different concentrations of ammonia from 5 to 100 ppm at room temperature [134]. Furthermore, a nonenzymatic electrochemical sensor based on 3D porous phase graphene oxide sheets encapsulated chalcopyrite (GOS@$CuFeS_2$) nanocomposite is reported for the detection of methyl paraoxon [135]. Encapsulation of nonconducting silicon oxide nanoparticles within conducting rGO can be used as the "bridging-material" in a field-effect transistor-based biosensor [130, 136]. Similarly, Si nanoparticles encapsulated with rGO via electrostatic interaction using APTES has also been reported [137, 138], resulting in less destruction and aggregation of SiNPs as compared to pristine nanoparticles. It also exhibited a high reversible capacity of 902 mAh g^{-1} after 100 cycles at 300 mA g^{-1} when used as the electrode.

4. Graphene oxide-based nanohybrids as pesticide biosensors

An analytical device that utilizes a biological sensing element to detect a specific analyte molecule or family of the analytical molecule is called as biosensor. Biosensors can seek applications in diverse fields such as food safety, environmental monitoring, and biomedical field. Generally, biosensors are consisting of two basic parts: first receptor, any organic or inorganic material that interacts with analytes. The second part, a transducer, which converts a recognition event, happened between analyte and receptor, into a measurable signal. Evaluation of biosensor 's performance is measured by its sensitivity to target, linear range, the limit of detection, dynamic ranges, reproducibility, precision in response, and selectivity [139]. Other parameters that are also important include the sensor's response time, ease of use, portability, storage, and operational stability. Graphene oxide/nanoparticle nanohybrids are known to be well suited for application in biosensing because of the rise of new advantageous properties due to the combination of graphene oxide and nanoparticles. Here, in this section, a detailed aspect of graphene oxide nanohybrid-based biosensors, specifically electrochemical biosensors are discussed.

Electrochemical sensors are the largest group of sensors for detecting or analyzing various molecules by directly converting biological recognition event into an electrical signal. A typical electrochemical biosensor is composed of a three-electrode system with a working electrode consisting of a biological recognition element, counter electrode, and reference electrode separated by suitable electrolytes. Based on their biological recognition process, electrochemical biosensors can be divided into two main groups: first, affinity-based sensors, and second, catalytic sensors. The basic principle of working in affinity biosensors is the measurable electric signal that arises due to the interaction of the biological component like an antibody, enzyme, nucleic acid, or a receptor and target molecules. Whereas in catalytic sensors, incorporated nanoparticles or enzymes recognize the analyte molecules and produce an electroactive species by catalysis. The electrical signal produced by the electroactive species is then correlated to the concentration of the target analyte molecule. Commonly used techniques in electrochemical biosensing include different forms of voltammetry (e.g., cyclic, linear sweep, differential, square wave, etc.) and amperometry [139].

The large surface area of graphene oxide nanohybrids is beneficial for the immobilization of biomolecules to use it as a platform for biosensing material. Furthermore, the synergistic effect of graphene oxide also enhances achievable sensitivities and measurable ranges. Most commonly biomolecule immobilized biosensors utilize enzymes, antibodies, and DNA as biomolecules.

Lately, enzyme immobilized GO nanohybrids-based biosensors have fascinated a lot for the detection of various kinds of analytes. The most common example is the determination of glucose, which has an important role in the diagnosis and therapy of diabetes. Apart from glucose oxidase based biosensors, other enzyme-based biosensors are also known with high sensitivity and selectivity, which includes biosensors based on alcohol dehydrogenase [140], microperoxidase [141], horseradish peroxidase [142], tyrosinase [143], urease [144], and acetylcholinesterase [145]. Acetylcholinesterase (AChE) is a catalytic enzyme present in the central nervous system, which catalyzes the hydrolysis of acetylcholine and choline esters. Its catalytic ability is severely affected by the presence of different types of organophosphorus and carbamate pesticides even in trace amounts. AChE can be easily immobilized on the surface of graphene oxide-based nanohybrids which offer a large surface area and abundant active sites so that they can be used for developing AChE inhibition-based biosensors [146].

Although a lot of work has already been reported on graphene-based biosensors; however, due to novel microbes and diseases associated with them, excess use of toxicants in food and feed products, nonjudicial use of pesticide and day by day disintegrating environmental conditions urgently need tools for detection of such chemicals and biologicals, and hence, more rapid and urgent requirement for the development of biosensors arises In the past 2 years, countless new graphene oxide nanohybrids-based biosensors are reported. For example, Yao et al. [147] reported an electrochemical biosensor based on the inhibition of AChE, using a gold nanocage/graphene oxide-chitosan nanocomposite-modified screen-printed carbon electrode for detection of chlorpyrifos (**Figure 2**). Where the biosensor showed good electrocatalytic activity for the oxidation of enzymatically produced thiocholine and detected chlorpyrifos concentrations as low as 3 ng L^{-1}.

Similarly, Bao et al. [148] developed a biosensor for malathion detection based on three-dimensional graphene-copper oxide nanoflowers nanocomposites electrode, and the group was able to obtain a wide linear relationship to malathion concentration ranging from 3 pM to 46.665 nM with a theoretical limit of detection at 0.92 pM. Moreover, Cui et al. reported a very stable electrochemical AChE biosensor for detection of dichlorvos by adsorption of AChE on chitosan, TiO_2 sol-gel,

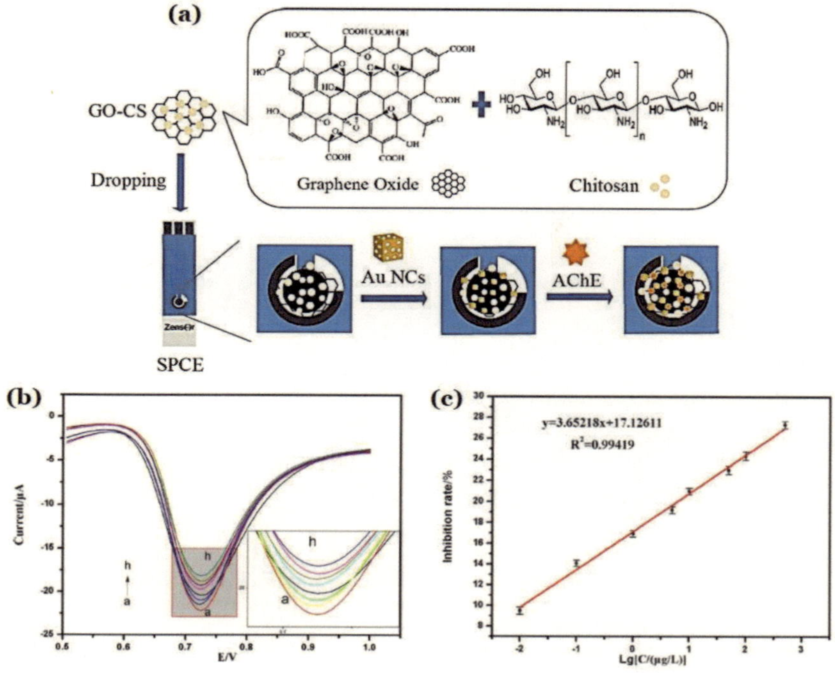

Figure 2.
(a) A schematic diagram of the construction process of AChE biosensor based on screen-printed electrodes, (b) and (c) DPV behavior, and percent inhibition obtained by biosensor in the presence of chlorpyrifos. Reprinted with permission from ref. [147], published by the Royal Society of Chemistry (RSC) on behalf of the Centre National de la Recherche Scientifique (CNRS) and the RSC.

and rGO-based many fold matrix, with the linear range varying from 0.036 μM to 22.6 μM, limit of detection of 29 nM and total time for detection about 25 min. Furthermore, electrochemical acetylcholinesterase biosensor based on the silver nanowire, graphene, TiO_2 sol-gel, chitosan, and acetylcholinesterase is fabricated by Zhang et al. [149] (**Figure 3**).

On a similar note, Zhang et al. [150] developed a highly sensitive AChE amperometric biosensor based on conjugated polymer and Ag-rGO-NH_2 nanocomposite. Group used a slightly different method for electrode fabrication where authors first electrochemically polymerized 4, 7-di (furan-2-yl) benzothiadiazole on electrode surface followed by deposition of Ag-rGO-NH_2 nanocomposite. The biosensor is found to be biocompatible with high efficiency having the linear range from 0.099 to 9.9 μg L^{-1} 0.032 μg L^{-1} for malathion and 0.001 μg L^{-1} for trichlorfon. Moreover, Mogha et al. [151] developed a biosensor for the detection of Chlorpyrifos using rGO supported Zirconium Oxide immobilized AChE (**Figure 4**). The group is able to detect the Chlorpyrifos in two linear ranges first from 10^{-13} to 10^{-9} M, whereas the second linear range was observed between 10^{-9} and 10^{-4} M.

Aghaie et al. [152] developed a nonenzymatic biosensor for the detection of paraoxon ethyl. A graphene-based NiFe bimetallic phosphosulfide nanocomposite biosensor is fabricated, where square wave voltammetric is used as a detection technique. The linear range for the detection of paraoxon methyl is found to be 12.3–10,000 nmol L^{-1} and limit of detection as 3.7 nmol L^{-1}.

Furthermore, a group of Hondred et al. [153] (**Figure 5**) utilized salt impregnated inkjet maskless lithography for preparation of 3D porous architectured graphene for application in biosensing of paraoxon and supercapacitor. The as developed biosensor showed a wide linear range from 10 to 500 nM, low limit of

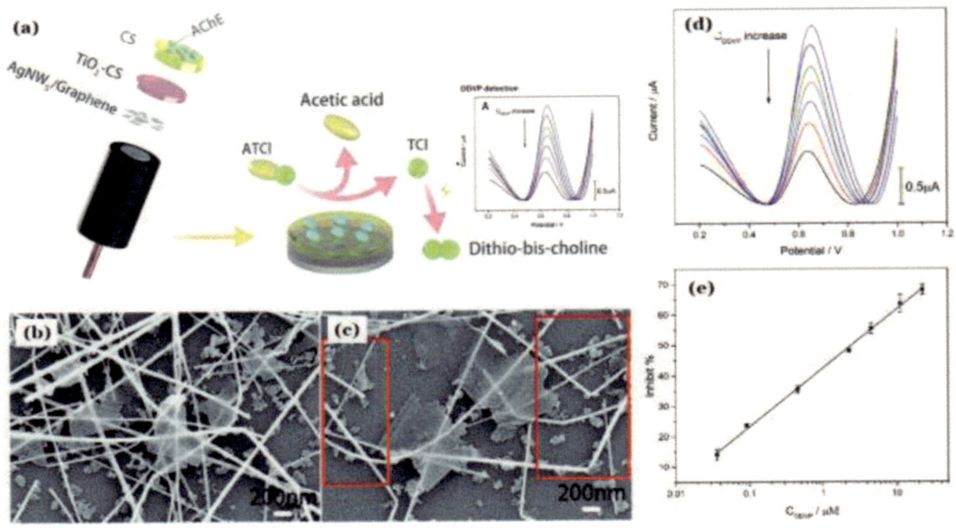

Figure 3.
(a) Schematic illustration of the AChE based biosensor and its working mechanism to ATCl. (b) and (c) SEM images of Gra/AgNWs/SiO$_2$ nanohybrids, where large graphene sheet is enhancing the connection with AgNWs in (b) with small graphene pieces further improving the performance of biosensor (in red rectangle shown in (c)), while (d) and (e) represent the inhibition of AChE in presence of DDVP using the biosensors. Reprinted with permission from ref [149], Published by The Royal Society of Chemistry.

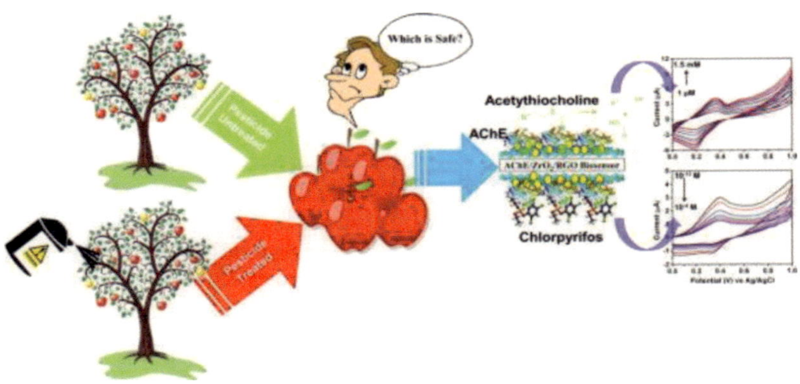

Figure 4.
An illustration of rGO supported Zirconium Oxide immobilized AChE nanohybrid as a biosensing platform for chlorpyrifos detection. Reprinted with permission from Ref. [151], published by Elsevier.

detection of 0.6 nM with high sensitivity of 12.4 nA nM^{-1}; moreover as a supercapacitor, it demonstrates a high energy density of 0.25 mW h cm^{-3} at a power density of 0.3 W cm^{-3}. Similarly, another AChE biosensor is developed based on a film of gold nanoparticles/three-dimensional graphene, by Dong et al. [154], for methyl parathion and malathion detection in a linear range from 1.0×10^{-10} to 1.0×10^{-6} g L^{-1}, having limits of detection as 2.78×10^{-11} g L^{-1} and 2.17×10^{-11} g L^{-1}.

Some more examples of biosensors based on graphene oxide nanohybrids for the detection of different types of pesticides such as methyl parathion [155–158], carbofuran [155, 157, 159], chlorpyrifos [156], imidacloprid [160], phoxim with graphene/GCE [161], poly(3-methylthiophene)/nitrogen-doped graphene [162], and carboxylic chitosan /silver nanoclusters-rGO [163], paraoxon and chlorpyrifos with TiO_2-GO/UiO-66 composite [164], carbaryl with MWCNTs/GO nanoribbons [165], carbaryl and chlorpyrifos with AgNPs-CGR/NF composite [166],

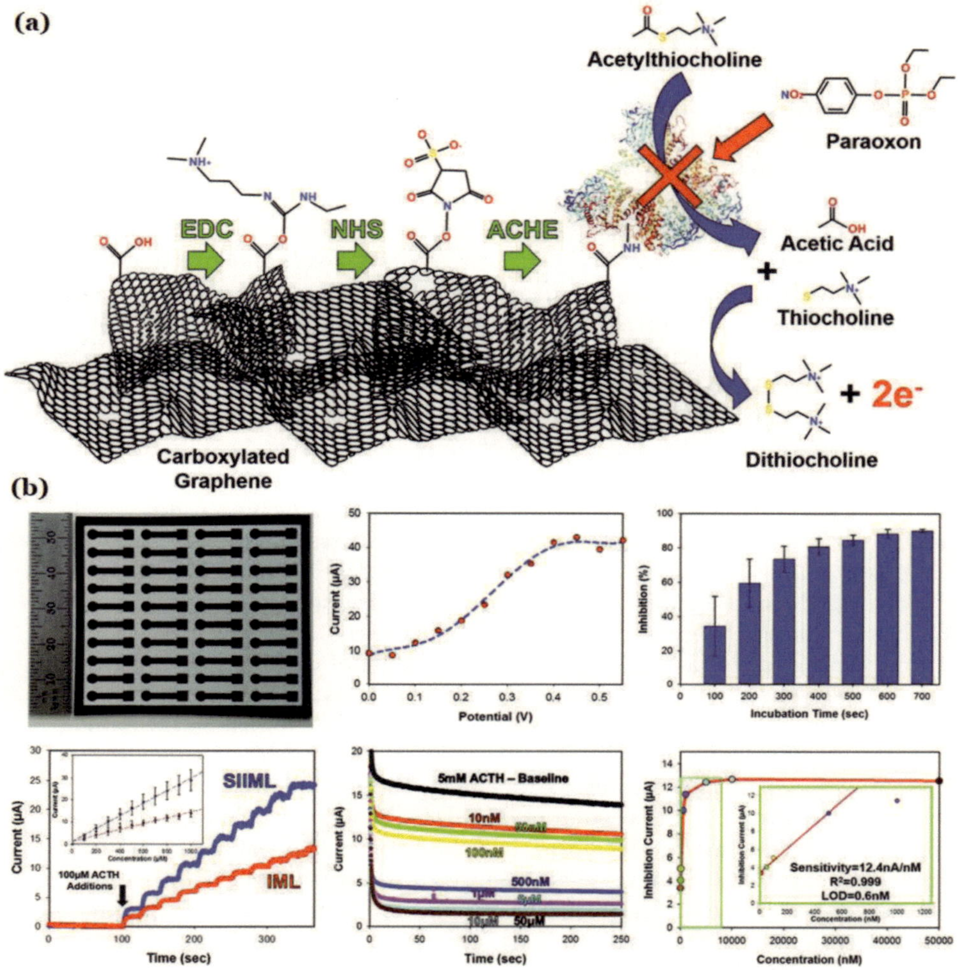

Figure 5.
(a) Schematic diagram of AChE biosensor portraying the functionalization approach for pesticide detection using EDC/NHS. (b) AChE pesticides biosensor characterization with photograph, activity, sensitivity, and comparison between salt impregnated inkjet maskless lithography (SIIML) and inkjet maskless lithography (IML)-based biosensors. Reprinted with permission from Ref. [153], published by The Royal Society of Chemistry.

chlorpyrifos and carbofuran with ZnONPs-CGR/NF composite [167], carbaryl and monocrotophos with ionic liquid-functionalized graphene /gelatin [168], monocrotophos with Prussian blue nanocubes [169], malathion and carbaryl with rGO-AuNP/β-cyclodextrin/Prussian blue-CS nanocomposites [170], fenitrothion with cerium oxide nanoparticle-decorated rGO [171], diuron with rGO-AuNPs [172], paraoxon-ethyl with rGO-AuNPs/polypyrrole [173], carbaryl with Graphene/polyaniline nanohybrid [174], carbaryl with an electrochemically induced porous GO network [175], and methyl parathion and malathion with plant esterase—Chit/AuNPs-graphene nanosheets [176].

5. Conclusion and future aspects

Graphene oxide is an attractive material that has gathered ever accumulative interest from the scientific community over the past several years. Owing to its extraordinary properties, graphene oxide and its derivatives are already being exploited in a wide variety of applications comprising electronics, energy, biosensors, catalysis, green chemistry, etc. Though, in the last decade, the relentless search for new opportunities benefiting from graphene oxide has led to the introduction and evolution of graphene oxide-based nanohybrids, which combine matchless and beneficial properties of nanomaterials/nanotechnology with those of graphene oxide to yield valuable and synergistic effects.

In this chapter, we have discussed the brief history of graphene oxide and graphene, emphasizing the structural details of graphene oxide and excellent properties associated with it. Graphene oxide-based nanohybrids show the synergistic effect of having properties of both graphene oxide as well as other constituting material whether nanoparticle or polymer. Synthetic mythologies of graphene oxide-based nanohybrids have also been discussed here in this chapter, in particular, graphene oxide/nanoparticle nanohybrids. Finally, applications of graphene oxide-based nanohybrids were presented in the field of biosensors and catalysis. In the case of biosensors, the main emphasis was given to the largest class of biosensors, that is, electrochemical biosensors, which consist of mainly enzyme biosensors and electrochemical DNA sensors, but some cases of other electrochemical sensors were also demonstrate. Applications of these graphene oxide-based hybrids in catalysis were also discussed emphasizing their use as an organic reaction catalyst, photocatalysts for the degradation of environmentally harmful molecules.

In conclusion, we have highlighted the properties of graphene oxide-based nanohybrids wherein these nanostructures can bring excellent synergistic advantages to a wide variety of biosensing applications. While promising, the field of graphene oxide-based nanohybrids is still not completely exhausted and several intriguing issues must be resolved before its maximum potential can be achieved. Besides, we envision that the evolution of this technology will result in the use of graphene oxide-based nanohybrids in a much wider range of applications by employing high quality and large-scale fabrication of these materials to minimize the cost leading to their commercialization. We also hope that this chapter has motivated attention from various disciplines that will gain benefits from the expansion of graphene oxide-based nanohybrids development for applications in numerous other fields of interest.

Acknowledgements

The author is heartily thankful to the Director, Shriram Institute for Industrial Research and Monica Singh, In-charge, Pesticide Laboratory for unconditional help and support.

Author details

Navin Kumar Mogha
Pesticide Laboratory, Analytical Science Division-Biology, Shriram Institute for Industrial Research, Delhi, India

*Address all correspondence to: navinmogha@gmail.com

References

[1] Bao LJ, Wei YL, Yao Y, Ruan QQ, Zeng EY. Global trends of research on emerging contaminants in the environment and humans: A literature assimilation. Environmental Science and Pollution Research. 2015;**22**:1635-1643. DOI: 10.1007/s11356-014-3404-8

[2] Gavrilescu M, Demnerová K, Aamand J, Agathos S, Fava F. Emerging pollutants in the environment: Present and future challenges in biomonitoring, ecological risks and bioremediation. New Biotechnology. 2015;**32**:147-156. DOI: 10.1016/j.nbt.2014.01.001

[3] Grandjean P, Landrigan PJ. Neurobehavioural effects of developmental toxicity. Lancet Neurology. 2014;**13**:330-338. DOI: 10.1016/S1474-4422(13)70278-3

[4] Moser F, Dondi F. Environmental protection between chemical practice and applied ethics: A critical review. Toxicological and Environmental Chemistry. 2016;**98**:1026-1042. DOI: 10.1080/02772248.2015.1025786

[5] Pundir CS, Malik A, Preety. Bio-sensing of organophosphorus pesticides: A review. Biosensors & Bioelectronics. 2019;**140**:111348. DOI: 10.1016/j.bios.2019.111348

[6] Liu M, Khan A, Wang Z, Liu Y, Yang G, Deng Y, et al. Aptasensors for pesticide detection. Biosensors & Bioelectronics. 2019;**130**:174-184. DOI: 10.1016/j.bios.2019.01.006

[7] Damalas CA, Koutroubas SD. Farmers' exposure to pesticides: Toxicity types and ways of prevention. Toxics. 2016;**4**:1. DOI: 10.3390/toxics4010001

[8] Ye M, Beach J, Martin JW, Senthilselvan A. Occupational pesticide exposures and respiratory health. International Journal of Environmental Research and Public Health. 2013;**10**:6442-6471. DOI: 10.3390/ijerph10126442

[9] Damalas CA, Eleftherohorinos IG. Pesticide exposure, safety issues, and risk assessment indicators. International Journal of Environmental Research and Public Health. 2011;**8**:1402-1419. DOI: 10.3390/ijerph8051402

[10] Kaur N, Prabhakar N. Current scenario in organophosphates detection using electrochemical biosensors. TrAC, Trends in Analytical Chemistry. 2017;**92**:62-85. DOI: 10.1016/j.trac.2017.04.012

[11] Xia N, Gao Y. Carbon nanostructures for development of acetylcholinesterase electrochemical biosensors for determination of pesticides. International Journal of Electrochemical Science. 2015;**10**:713-724

[12] Boehm HP, Setton R, Stumpp E. Nomenclature and terminology of graphite intercalation compounds (IUPAC recommendations 1994). Pure and Applied Chemistry. 1994;**66**:1893-1901. DOI: 10.1351/pac199466091893

[13] Novoselov KS, Geim AK, Morozov SV, Jiang D, Zhang Y, Dubonos SV, et al. Electric field effect in atomically thin carbon films. Science. 2004;**306**:666-669. DOI: 10.1126/science.1102896

[14] Novoselov KS, Geim AK, Morozov SV, Jiang D, Katsnelson MI, Grigorieva IV, et al. Two-dimensional gas of massless dirac fermions in graphene. Nature. 2005;**438**:197-200. DOI: 10.1038/nature04233

[15] Lau CN, Bao W, Velasco J. Properties of suspended graphene membranes. Materials Today. 2012;**15**:238-245. DOI: 10.1016/S1369-7021(12)70114-1

[16] Ruoff R. Calling all chemists. Nature Nanotechnology. 2008;**3**:10-11. DOI: 10.1038/nnano.2007.432

[17] Dreyer DR, Park S, Bielawski CW, Ruoff RS. The chemistry of graphene oxide. Chemical Society Reviews. 2010;**39**:228-240. DOI: 10.1039/B917103G

[18] Lerf A, He H, Forster M, Klinowski J. Structure of graphite oxide revisited ||. The Journal of Physical Chemistry. B. 1998;**102**:4477-4482. DOI: 10.1021/jp9731821

[19] Szabó T, Berkesi O, Forgó P, Josepovits K, Sanakis Y, Petridis D, et al. Evolution of surface functional groups in a series of progressively oxidized graphite oxides evolution of surface functional groups in a series of progressively oxidized graphite oxides. Chemistry of Materials. 2006;**18**:2740-2749. DOI: 10.1021/cm060258

[20] Gao W, Alemany LB, Ci L, Ajayan PM. New insights into the structure and reduction of graphite oxide. Nature Chemistry. 2009;**1**: 403-408. DOI: 10.1038/nchem.281

[21] Eda G, Chhowalla M. Chemically derived Graphene oxide: Towards large-area thin-film electronics and optoelectronics. Advanced Materials. 2010;**22**:2392-2415. DOI: 10.1002/adma.200903689

[22] Li X, Zhang G, Bai X, Sun X, Wang X, Wang E, et al. Highly conducting graphene sheets and Langmuir–Blodgett films. Nature Nanotechnology. 2008;**3**:538-542. DOI: 10.1038/nnano.2008.210

[23] Kim F, Cote LJ, Huang J. Graphene oxide: Surface activity and two-dimensional assembly. Advanced Materials. 2010;**22**:1954-1958. DOI: 10.1002/adma.200903932

[24] Compton OC, Nguyen ST. Graphene oxide, highly reduced graphene oxide, and Graphene: Versatile building blocks for carbon-based materials. Small. 2010;**6**:711-723. DOI: 10.1002/smll.200901934

[25] Filip J, Tkac J. Is graphene worth using in biofuel cells? Electrochimica Acta. 2014;**136**:340-354. DOI: 10.1016/j.electacta.2014.05.119

[26] Gómez-Navarro C, Meyer JC, Sundaram RS, Chuvilin A, Kurasch S, Burghard M, et al. Atomic structure of reduced graphene oxide. Nano Letters. 2010;**10**:1144-1148. DOI: 10.1021/nl9031617

[27] Mkhoyan KA, Contryman AW, Silcox J, Stewart DA, Eda G, Mattevi C, et al. Atomic and electronic structure of graphene-oxide. Nano Letters. 2009;**9**:1058-1063. DOI: 10.1021/nl8034256

[28] Bissessur R, Scully SF. Intercalation of solid polymer electrolytes into graphite oxide. Solid State Ionics. 2007;**178**:877-882. DOI: 10.1016/j.ssi.2007.02.030

[29] Paredes JI, Villar-Rodil S, Martínez-Alonso A, Tascón JMD. Graphene oxide dispersions in organic solvents. Langmuir. 2008;**24**:10560-10564. DOI: 10.1021/la801744a

[30] Mattevi C, Eda G, Agnoli S, Miller S, Mkhoyan KA, Celik O, et al. Evolution of electrical, chemical, and structural properties of transparent and conducting chemically derived graphene thin films. Advanced Functional Materials. 2009;**19**:2577-2583. DOI: 10.1002/adfm.200900166

[31] Eda G, Fanchini G, Chhowalla M. Large-area ultrathin films of reduced graphene oxide as a transparent and flexible electronic material. Nature Nanotechnology. 2008;**3**:270-274. DOI: 10.1038/nnano.2008.83

[32] Eda G, Mattevi C, Yamaguchi H, Kim H, Chhowalla M. Insulator to semimetal transition in graphene oxide. Journal of Physical Chemistry C. 2009;**113**:15768-15771. DOI: 10.1021/jp9051402

[33] Gómez-Navarro C, Weitz RT, Bittner AM, Scolari M, Mews A, Burghard M, et al. Electronic transport properties of individual chemically reduced graphene oxide sheets. Nano Letters. 2007;**7**:3499-3503. DOI: 10.1021/nl072090c

[34] Schniepp HC, Li JL, McAllister MJ, Sai H, Herrera-Alonson M, Adamson DH, et al. Functionalized single graphene sheets derived from splitting graphite oxide. The Journal of Physical Chemistry. B. 2006;**110**:8535-8539. DOI: 10.1021/jp060936f

[35] Stankovich S, Dikin DA, Piner RD, Kohlhaas KA, Kleinhammes A, Jia Y, et al. Synthesis of graphene-based nanosheets via chemical reduction of exfoliated graphite oxide. Carbon N Y. 2007;**45**:1558-1565. DOI: 10.1016/j.carbon.2007.02.034

[36] Cote LJ, Kim F, Huang J. Langmuir-Blodgett assembly of graphite oxide single layers. Journal of the American Chemical Society. 2009;**131**:1043-1049. DOI: 10.1021/ja806262m

[37] Paredes JI, Villar-Rodil S, Solís-Fernández P, Martínez-Alonso A, Tascón JMD. Atomic force and scanning tunneling microscopy imaging of graphene nanosheets derived from graphite oxide. Langmuir. 2009;**25**:5957-5968. DOI: 10.1021/la804216z

[38] Mativetsky JM, Treossi E, Orgiu E, Melucci M, Veronese GP, Samorì P, et al. Local current mapping and patterning of reduced graphene oxide. Journal of the American Chemical Society. 2010;**132**:14130-14136. DOI: 10.1021/ja104567f

[39] Erickson K, Erni R, Lee Z, Alem N, Gannett W, Zettl A. Determination of the local chemical structure of graphene oxide and reduced graphene oxide. Advanced Materials. 2010;**22**:4467-4472. DOI: 10.1002/adma.201000732

[40] Pantelic RS, Meyer JC, Kaiser U, Baumeister W, Plitzko JM. Graphene oxide: A substrate for optimizing preparations of frozen-hydrated samples. Journal of Structural Biology. 2010;**170**:152-156. DOI: 10.1016/j.jsb.2009.12.020

[41] Pacilé D, Meyer JC, Fraile Rodríguez A, Papagno M, Gómez-Navarro C, Sundaram RS, et al. Electronic properties and atomic structure of graphene oxide membranes. Carbon N Y. 2011;**49**:966-972. DOI: 10.1016/j.carbon.2010.09.063

[42] Casabianca LB, Shaibat MA, Cai WW, Park S, Piner R, Ruoff RS, et al. NMR-based structural modeling of graphite oxide using multidimensional 13C solid-state NMR and ab initio chemical shift calculations. Journal of the American Chemical Society. 2010;**132**:5672-5676. DOI: 10.1021/ja9030243

[43] Si Y, Samulski ET. Synthesis of water soluble graphene. Nano Letters. 2008;**8**:1679-1682. DOI: 10.1021/nl080604h

[44] Cai W, Piner RD, Stadermann FJ, Park S, Shaibat MA, Ishii Y, et al. Synthesis and solid-state NMR structural characterization of 13C-labeled graphite oxide. Science. 2008;**321**:1815-1817. DOI: 10.1126/science.1162369

[45] Lee V, Whittaker L, Jaye C, Baroudi KM, Fischer DA, Banerjee S. Large-area chemically medified graphene films: Electrophoretic deposition and characterization by soft X-ray absorption spectroscopy. Chemistry

of Materials. 2009;**21**:3905-3916. DOI: 10.1021/cm901554p

[46] Saxena S, Tyson TA, Negusse E. Investigation of the local structure of graphene oxide. Journal of Physical Chemistry Letters. 2010;**1**:3433-3437. DOI: 10.1021/jz1014339

[47] Lee DW, De Los Santos VL, Seo JW, Felix LL, Bustamante DA, Cole JM, et al. The structure of graphite oxide: Investigation of its surface chemical groups. The Journal of Physical Chemistry. B. 2010;**114**:5723-5728. DOI: 10.1021/jp1002275

[48] Saxena S, Tyson TA, Shukla S, Negusse E, Chen H, Bai J. Investigation of structural and electronic properties of graphene oxide. Applied Physics Letters. 2011;**99**:013104. DOI: 10.1063/1.3607305

[49] Ganguly A, Sharma S, Papakonstantinou P, Hamilton J. Probing the thermal deoxygenation of graphene oxide using high-resolution in situ X-ray-based spectroscopies. Journal of Physical Chemistry C. 2011;**115**:17009-17019. DOI: 10.1021/jp203741y

[50] Bagri A, Mattevi C, Acik M, Chabal YJ, Chhowalla M, Shenoy VB. Structural evolution during the reduction of chemically derived graphene oxide. Nature Chemistry. 2010;**2**:581-587. DOI: 10.1038/nchem.686

[51] Acik M, Lee G, Mattevi C, Pirkle A, Wallace RM, Chhowalla M, et al. The role of oxygen during thermal reduction of graphene oxide studied by infrared absorption spectroscopy. Journal of Physical Chemistry C. 2011;**115**:19761-19781. DOI: 10.1021/jp2052618

[52] Yang D, Velamakanni A, Bozoklu G, Park S, Stoller M, Piner RD, et al. Chemical analysis of graphene oxide films after heat and chemical treatments by X-ray photoelectron and micro-Raman spectroscopy. Carbon N Y. 2009;**47**:145-152. DOI: 10.1016/j.carbon.2008.09.045

[53] Pimenta MA, Dresselhaus G, Dresselhaus MS, Cançado LG, Jorio A, Saito R. Studying disorder in graphite-based systems by Raman spectroscopy. Physical Chemistry Chemical Physics. 2007;**9**:1276-1291. DOI: 10.1039/b613962k

[54] Dresselhaus MS, Dresselhaus G, Hofmann M. Raman spectroscopy as a probe of graphene and carbon nanotubes. Philosophical Transactions of the Royal Society A - Mathematical Physical and Engineering Sciences. 2008;**366**:231-236. DOI: 10.1098/rsta.2007.2155

[55] Boukhvalov DW, Katsnelson MI. Modeling of graphite oxide. Journal of the American Chemical Society. 2008;**130**:10697-10701. DOI: 10.1021/ja8021686

[56] Becerril HA, Mao J, Liu Z, Stoltenberg RM, Bao Z, Chen Y. Evaluation of solution-processed reduced graphene oxide films as transparent conductors. ACS Nano. 2008;**2**:463-470. DOI: 10.1021/nn700375n

[57] Wang S, Chia PJ, Chua LL, Zhao LH, Png RQ, Sivaramakrishnan S, et al. Band-like transport in surface-functionalized highly solution-processable graphene nanosheets. Advanced Materials. 2008;**20**:3440-3446. DOI: 10.1002/adma.200800279

[58] Du X, Skachko I, Barker A, Andrei EY. Approaching ballistic transport in suspended graphene. Nature Nanotechnology. 2008;**3**:491-495. DOI: 10.1038/nnano.2008.199

[59] López V, Sundaram RS, Gómez-Navarro C, Olea D, Burghard M, Gómez-Herrero J, et al. Chemical vapor deposition repair of graphene oxide:

A route to highly conductive graphene monolayers. Advanced Materials. 2009;**21**:4683-4686. DOI: 10.1002/adma.200901582

[60] Lee G, Kim KS, Cho K. Theoretical study of the electron transport in graphene with vacancy and residual oxygen defects after high-temperature reduction. Journal of Physical Chemistry C. 2011;**115**:9719-9725. DOI: 10.1021/jp111841w

[61] Li D, Müller MB, Gilje S, Kaner RB, Wallace GG. Processable aqueous dispersions of graphene nanosheets. Nature Nanotechnology. 2008;**3**:101-105. DOI: 10.1038/nnano.2007.451

[62] Jung I, Field DA, Clark NJ, Zhu Y, Yang D, Piner RD, et al. Reduction kinetics of graphene oxide determined by electrical transport measurements and temperature programmed desorption. Journal of Physical Chemistry C. 2009;**113**:18480-18486. DOI: 10.1021/jp904396j

[63] Loh KP, Bao Q, Eda G, Chhowalla M. Graphene oxide as a chemically tunable platform for optical applications. Nature Chemistry. 2010;**2**:1015-1024. DOI: 10.1038/nchem.907

[64] Eda G, Lin YY, Mattevi C, Yamaguchi H, Chen HA, Chen IS, et al. Blue photoluminescence from chemically derived graphene oxide. Advanced Materials. 2010;**22**:505-509. DOI: 10.1002/adma.200901996

[65] Liu F, Choi JY, Seo TS. Graphene oxide arrays for detecting specific DNA hybridization by fluorescence resonance energy transfer. Biosensors & Bioelectronics. 2010;**25**:2361-2365. DOI: 10.1016/j.bios.2010.02.022

[66] Tang L, Wang Y, Li Y, Feng H, Lu J, Li J. Preparation, structure, and electrochemical properties of reduced graphene sheet films. Advanced Functional Materials. 2009;**19**:2782-2789. DOI: 10.1002/adfm.200900377

[67] Wang Y, Wan Y, Zhang D. Reduced graphene sheets modified glassy carbon electrode for electrocatalytic oxidation of hydrazine in alkaline media. Electrochemistry Communications. 2010;**12**:187-190. DOI: 10.1016/j.elecom.2009.11.019

[68] Shao Y, Wang J, Engelhard M, Wang C, Lin Y. Facile and controllable electrochemical reduction of graphene oxide and its applications. Journal of Materials Chemistry. 2010;**20**:743-748. DOI: 10.1039/b917975e

[69] Wang DW, Li F, Wu ZS, Ren W, Cheng HM. Electrochemical interfacial capacitance in multilayer graphene sheets: Dependence on number of stacking layers. Electrochemistry Communications. 2009;**11**:1729-1732. DOI: 10.1016/j.elecom.2009.06.034

[70] Park S, An J, Potts JR, Velamakanni A, Murali S, Ruoff RS. Hydrazine-reduction of graphite- and graphene oxide. Carbon N Y. 2011;**49**:3019-3023. DOI: 10.1016/j.carbon.2011.02.071

[71] Shin HJ, Kim KK, Benayad A, Yoon SM, Park HK, Jung IS, et al. Efficient reduction of graphite oxide by sodium borohydride and its effect on electrical conductance. Advanced Functional Materials. 2009;**19**:1987-1992. DOI: 10.1002/adfm.200900167

[72] Guoxiu W, Juan Y, Jinsoo P, Xinglong G, Bei W, Hao L, et al. Facile synthesis and characterization of graphene nanosheets. Journal of Physical Chemistry C. 2008;**112**:8192-8195. DOI: 10.1021/jp710931h

[73] Ramesha GK, Sampath NS. Electrochemical reduction of oriented graphene oxide films: An in situ Raman spectroelectrochemical study. Journal of

Physical Chemistry C. 2009;**113**:7985-7989. DOI: 10.1021/jp811377n

[74] Niyogi S, Bekyarova E, Itkis ME, McWilliams JL, Hamon MA, Haddon RC. Solution properties of graphite and graphene. Journal of the American Chemical Society. 2006;**128**:7720-7721. DOI: 10.1021/ja060680r

[75] Liu ZB, Xu YF, Zhang XY, Zhang XL, Chen YS, Tian JG. Porphyrin and fullerene covalently functionalized graphene hybrid materials with large nonlinear optical properties. The Journal of Physical Chemistry. B. 2009;**113**:9681-9686. DOI: 10.1021/jp9004357

[76] Fang M, Wang K, Lu H, Yang Y, Nutt S. Covalent polymer functionalization of graphene nanosheets and mechanical properties of composites. Journal of Materials Chemistry. 2009;**19**:7098-7105. DOI: 10.1039/b908220d

[77] Yang H, Shan C, Li F, Han D, Zhang Q, Niu L. Covalent functionalization of polydisperse chemically-converted graphene sheets with amine-terminated ionic liquid. Chemical Communications. 2009:3880-3882. DOI: 10.1039/b905085j

[78] Yang X, Zhang X, Liu Z, Ma Y, Huang Y, Chen Y. High-efficiency loading and controlled release of doxorubicin hydrochloride on graphene oxide. Journal of Physical Chemistry C. 2008;**112**:17554-17558. DOI: 10.1021/jp806751k

[79] Lu CH, Yang HH, Zhu CL, Chen X, Chen GN. A graphene platform for sensing biomolecules. Angewandte Chemie, International Edition. 2009;**48**:4785-4787. DOI: 10.1002/anie.200901479

[80] Dreyer DR, Jia HP, Bielawski CW. Graphene oxide: A convenient carbocatalyst for facilitating oxidation and hydration reactions. Angewandte Chemie, International Edition. 2010;**49**:6813-6816. DOI: 10.1002/anie.201002160

[81] Dreyer DR, Jarvis KA, Ferreira PJ, Bielawski CW. Graphite oxide as a dehydrative polymerization catalyst: A one-step synthesis of carbon-reinforced poly(phenylene methylene) composites. Macromolecules. 2011;**44**:7659-7667. DOI: 10.1021/ma201306x

[82] Dreyer DR, Jarvis KA, Ferreira PJ, Bielawski CW. Graphite oxide as a carbocatalyst for the preparation of fullerene-reinforced polyester and polyamide nanocomposites. Polymer Chemistry. 2012;**3**:757-766. DOI: 10.1039/c2py00545j

[83] Jia HP, Dreyer DR, Bielawski CW. C-H oxidation using graphite oxide. Tetrahedron. 2011;**67**:4431-4434. DOI: 10.1016/j.tet.2011.02.065

[84] Shah B, Yin PT, Ghoshal S, Lee KB. Multimodal magnetic core-shell nanoparticles for effective stem-cell differentiation and imaging. Angewandte Chemie, International Edition. 2013;**52**:6190-6195. DOI: 10.1002/anie.201302245

[85] Tjoa V, Jun W, Dravid V, Mhaisalkar S, Mathews N. Hybrid graphene-metal nanoparticle systems: Electronic properties and gas interaction. Journal of Materials Chemistry. 2011;**21**:15593-15599. DOI: 10.1039/c1jm12676h

[86] Artiles MS, Rout CS, Fisher TS. Graphene-based hybrid materials and devices for biosensing. Advanced Drug Delivery Reviews. 2011;**63**:1352-1360. DOI: 10.1016/j.addr.2011.07.005

[87] Mogha NK, Gosain S, Masram DT. Gold nanoworms immobilized graphene oxide polymer brush nanohybrid for catalytic degradation studies of

organic dyes. Applied Surface Science. 2017;**396**:1427-1434. DOI: 10.1016/j.apsusc.2016.11.182

[88] Mogha NK, Gosain S, Masram DT. Lanthanum oxide nanoparticles immobilized reduced graphene oxide polymer brush nanohybrid for environmental vitiation of organic dyes. Arabian Journal of Chemistry. 2020;**13**:1367-1376. DOI: 10.1016/j.arabjc.2017.11.008

[89] Zhang M, Li Y, Su Z, Wei G. Recent advances in the synthesis and applications of graphene-polymer nanocomposites. Polymer Chemistry. 2015;**6**:6107-6124. DOI: 10.1039/c5py00777a

[90] Ye YS, Chen YN, Wang JS, Rick J, Huang YJ, Chang FC, et al. Versatile grafting approaches to functionalizing individually dispersed graphene nanosheets using RAFT polymerization and click chemistry. Chemistry of Materials. 2012;**24**:2987-2997. DOI: 10.1021/cm301345r

[91] Chen Y, Zhang S, Liu X, Pei Q, Qian J, Zhuang Q, et al. Preparation of solution-processable reduced graphene oxide/polybenzoxazole nanocomposites with improved dielectric properties. Macromolecules. 2015;**48**:365-372. DOI: 10.1021/ma502326v

[92] Yuan B, Bao C, Song L, Hong N, Liew KM, Hu Y. Preparation of functionalized graphene oxide/polypropylene nanocomposite with significantly improved thermal stability and studies on the crystallization behavior and mechanical properties. Chemical Engineering Journal. 2014;**237**:411-420. DOI: 10.1016/j.cej.2013.10.030

[93] Hu H, Zhang G, Xiao L, Wang H, Zhang Q, Zhao Z. Preparation and electrical conductivity of graphene/ultrahigh molecular weight polyethylene composites with a segregated structure. Carbon N Y. 2012;**50**:4596-4599. DOI: 10.1016/j.carbon.2012.05.045

[94] Kong BS, Geng J, Jung HT. Layer-by-layer assembly of graphene and gold nanoparticles by vacuum filtration and spontaneous reduction of gold ions. Chemical Communications. 2009:2174-2176. DOI: 10.1039/b821920f

[95] Hong W, Bai H, Xu Y, Yao Z, Gu Z, Shi G. Preparation of gold nanoparticle/graphene composites with controlled weight contents and their application in biosensors. Journal of Physical Chemistry C. 2010;**114**:1822-1826. DOI: 10.1021/jp9101724

[96] Tien HW, Huang YL, Yang SY, Wang JY, Ma CCM. The production of graphene nanosheets decorated with silver nanoparticles for use in transparent, conductive films. Carbon N Y. 2011;**49**:1550-1560. DOI: 10.1016/j.carbon.2010.12.022

[97] Qiu JD, Wang GC, Liang RP, Xia XH, Yu HW. Controllable deposition of platinum nanoparticles on graphene as an electrocatalyst for direct methanol fuel cells. Journal of Physical Chemistry C. 2011;**115**:15639-15645. DOI: 10.1021/jp200580u

[98] Lin J, Raji ARO, Nan K, Peng Z, Yan Z, Samuel ELG, et al. Iron oxide nanoparticle and graphene nanoribbon composite as an anode material for high-performance li-ion batteries. Advanced Functional Materials. 2014;**24**:2044-2048. DOI: 10.1002/adfm.201303023

[99] Liang Y, Wang H, Casalongue HS, Chen Z, Dai H. TiO_2 Nanocrystals grown on graphene as advanced photocatalytic hybrid materials. Nano Research. 2010;**3**:701-705. DOI: 10.1007/s12274-010-0033-5

[100] Son DI, Kwon BW, Park DH, Seo WS, Yi Y, Angadi B, et al. Emissive ZnO-graphene quantum dots for

white-light-emitting diodes. Nature Nanotechnology. 2012;7:465-471. DOI: 10.1038/nnano.2012.71

[101] Lin J, Peng Z, Xiang C, Ruan G, Yan Z, Natelson D, et al. Graphene nanoribbon and nanostructured SnO_2 composite anodes for lithium ion batteries. ACS Nano. 2013;7:6001-6006. DOI: 10.1021/nn4016899

[102] Deng S, Tjoa V, Fan HM, Tan HR, Sayle DC, Olivo M, et al. Reduced graphene oxide conjugated Cu_2O nanowire mesocrystals for high-performance NO_2 gas sensor. Journal of the American Chemical Society. 2012;**134**:4905-4917. DOI: 10.1021/ja211683m

[103] Subodh MNK, Chaudhary K, Kumar G, Masram DT. Fur-imine-functionalized graphene oxide-immobilized copper oxide nanoparticle catalyst for the synthesis of xanthene derivatives. ACS Omega. 2018;**3**:16377-16385. DOI: 10.1021/acsomega.8b01781

[104] He Y, Chen W, Li X, Zhang Z, Fu J, Zhao C, et al. Freestanding three-dimensional graphene/MnO_2 composite networks as ultralight and flexible supercapacitor electrodes. ACS Nano. 2013;7:174-182. DOI: 10.1021/nn304833s

[105] Zhou G, Wang DW, Yin LC, Li N, Li F, Cheng HM. Oxygen bridges between nio nanosheets and graphene for improvement of lithium storage. ACS Nano. 2012;**6**:3214-3223. DOI: 10.1021/nn300098m

[106] Kumar G, Mogha NK, Kumar M, Subodh MDT. NiO nanocomposites/rGO as a heterogeneous catalyst for imidazole scaffolds with applications in inhibiting the DNA binding activity. Dalton Transactions. 2020;**49**:1963-1974. DOI: 10.1039/c9dt04416g

[107] Mogha NK, Kirti S, Masram DT. La_2O_3/reduced graphene oxide nanocomposite: A highly efficient, reusable heterogeneous catalyst for the synthesis of biologically important bis(indolyl)methanes under solvent free conditions. Journal of Nanoscience and Nanotechnology. 2017;**17**:2508-2514. DOI: 10.1166/jnn.2017.13441

[108] Mogha NK, Sahu V, Sharma M, Sharma RK, Masram DT. Sensitive and reliable ascorbic acid sensing by lanthanum oxide/reduced graphene oxide nanocomposite. Applied Biochemistry and Biotechnology. 2014;**174**:1010-1020. DOI: 10.1007/s12010-014-0986-z

[109] Chen Y, Li Y, Sun D, Tian D, Zhang J, Zhu JJ. Fabrication of gold nanoparticles on bilayer graphene for glucose electrochemical biosensing. Journal of Materials Chemistry. 2011;**21**:7604-7611. DOI: 10.1039/c1jm10293a

[110] Jain PK, Huang X, El-Sayed IH, El-Sayed MA. Noble metals on the nanoscale: Optical and photothermal properties and some applications in imaging, sensing, biology, and medicine. Accounts of Chemical Research. 2008;**41**:1578-1586. DOI: 10.1021/ar7002804

[111] Eustis S, El-Sayed MA. Why gold nanoparticles are more precious than pretty gold: Noble metal surface plasmon resonance and its enhancement of the radiative and nonradiative properties of nanocrystals of different shapes. Chemical Society Reviews. 2006;**35**:209-217. DOI: 10.1039/b514191e

[112] Naqvi TK, Srivastava AK, Kulkarni MM, Siddiqui AM, Dwivedi PK. Silver nanoparticles decorated reduced graphene oxide (rGO) SERS sensor for multiple analytes. Applied Surface Science. 2019;**478**:887-895. DOI: 10.1016/j.apsusc.2019.02.026

[113] Zhang H, Hines D, Akins DL. Synthesis of a nanocomposite composed

of reduced graphene oxide and gold nanoparticles. Dalton Transactions. 2014;**43**:2670-2675. DOI: 10.1039/c3dt52573b

[114] Darabdhara G, Das MR, Singh SP, Rengan AK, Szunerits S, Boukherroub R. Ag and Au nanoparticles/reduced graphene oxide composite materials: Synthesis and application in diagnostics and therapeutics. Advances in Colloid and Interface Science. 2019;**271**:101991. DOI: 10.1016/j.cis.2019.101991

[115] Zhou Y, Bao Q, Tang LAL, Zhong Y, Loh KP. Hydrothermal dehydration for the "green" reduction of exfoliated graphene oxide to graphene and demonstration of tunable optical limiting properties. Chemistry of Materials. 2009;**21**:2950-2956. DOI: 10.1021/cm9006603

[116] Liu WM, Li J, Zhang HY. Reduced graphene oxide modified zinc oxide composites synergistic photocatalytic activity under visible light irradiation. Optik (Stuttg). 2020;**207**:163778. DOI: 10.1016/j.ijleo.2019.163778

[117] Kusiak-Nejman E, Wanag A, Kapica-Kozar J, Kowalczyk Ł, Zgrzebnicki M, Tryba B, et al. Methylene blue decomposition on TiO_2/reduced graphene oxide hybrid photocatalysts obtained by a two-step hydrothermal and calcination synthesis. Catalysis Today. 2019. DOI: 10.1016/j.cattod.2019.04.078

[118] Frindy S, Sillanpää M. Synthesis and application of novel α-Fe_2O_3/graphene for visible-light enhanced photocatalytic degradation of RhB. Materials and Design. 2020;**188**:108461. DOI: 10.1016/j.matdes.2019.108461

[119] Hu J, Chen M, Rong Q, Zhang Y, Wang H, Zhang D, et al. Formaldehyde sensing performance of reduced graphene oxide-wrapped hollow SnO_2 nanospheres composites. Sensors and Actuators B: Chemical. 2020;**307**:127584. DOI: 10.1016/j.snb.2019.127584

[120] Ren L, Huang S, Fan W, Liu T. One-step preparation of hierarchical superparamagnetic iron oxide/graphene composites via hydrothermal method. Applied Surface Science. 2011;**258**:1132-1138. DOI: 10.1016/j.apsusc.2011.09.049

[121] Indhumathy M, Prakasam A. A one pot hydrothermal stimulated CdS-reduced graphene oxide (CdS/rGO) hybrid nanocomposite for high performance photovoltaic applications. Journal of Materials Science: Materials in Electronics. 2019;**30**:15444-15451. DOI: 10.1007/s10854-019-01920-x

[122] Wang SX, Maimaiti H, Xu B, Awati A, Bin ZG, Dan CY. Synthesis and visible-light photocatalytic N_2/H_2O to ammonia of ZnS nanoparticles supported by petroleum pitch-based graphene oxide. Applied Surface Science. 2019;**493**:514-524. DOI: 10.1016/j.apsusc.2019.06.287

[123] Zhang X, Duan L, Zhang X, Li X, Lü W. Preparation of Cu2S@rGO hybrid composites as anode materials for enhanced electrochemical properties of lithium ion battery. Journal of Alloys and Compounds. 2020;**816**:152539. DOI: 10.1016/j.jallcom.2019.152539

[124] Hou X, Wang Z, Fan G, Ji H, Yi S, Li T, et al. Hierarchical three-dimensional MOS_2/GO hybrid nanostructures for triethylamine-sensing applications with high sensitivity and selectivity. Sensors and Actuators B: Chemical. 2020;**317**:128236. DOI: 10.1016/j.snb.2020.128236

[125] Bijani S, Martínez L, Gabás M, Dalchiele EA, Ramos-Barrado JR. Low-temperature electrodeposition of Cu_2O thin films: Modulation of micro-nanostructure by modifying the applied potential and electrolytic bath pH. Journal of Physical Chemistry C.

2009;**113**:19482-19487. DOI: 10.1021/jp905952a

[126] Zhao G, Liu G. Electrochemical deposition of gold nanoparticles on reduced graphene oxide by fast scan cyclic voltammetry for the sensitive determination of As(III). Nanomaterials. 2019;**9**:41. DOI: 10.3390/nano9010041

[127] Azevedo VHR, da Silva JL, Stradiotto NR. Silver oxide nanoparticles in reduced graphene oxide modified electrode for amino acids electrocatalytic oxidation. Journal of Electroanalytical Chemistry. 2019;**845**:57-65. DOI: 10.1016/j.jelechem.2019.05.037

[128] Saravanan G, Mohan S. Pt nanoparticles embedded on reduced graphite oxide with excellent electrocatalytic properties. Applied Surface Science. 2016;**386**:96-102. DOI: 10.1016/j.apsusc.2016.05.152

[129] Wu S, Yin Z, He Q, Lu G, Zhou X, Zhang H. Electrochemical deposition of Cl-doped n-type Cu_2O on reduced graphene oxide electrodes. Journal of Materials Chemistry. 2011;**21**:3467-3470. DOI: 10.1039/c0jm02267e

[130] Myung S, Solanki A, Kim C, Park J, Kim KS, Lee KB. Graphene-encapsulated nanoparticle-based biosensor for the selective detection of cancer biomarkers. Advanced Materials. 2011;**23**:2221-2225. DOI: 10.1002/adma.201100014

[131] Yang S, Feng X, Ivanovici S, Müllen K. Fabrication of graphene-encapsulated oxide nanoparticles: Towards high-performance anode materials for lithium storage. Angewandte Chemie, International Edition. 2010;**49**:8408-8411. DOI: 10.1002/anie.201003485

[132] Liu J, Li Y, Li Y, Li J, Deng Z. Noncovalent DNA decorations of graphene oxide and reduced graphene oxide toward water-soluble metal-carbon hybrid nanostructures via self-assembly. Journal of Materials Chemistry. 2010;**20**:900-906. DOI: 10.1039/b917752c

[133] Lee JK, Smith KB, Hayner CM, Kung HH. Silicon nanoparticles-graphene paper composites for Li ion battery anodes. Chemical Communications. 2010;**46**:2025-2027. DOI: 10.1039/b919738a

[134] Feng Q, Li X, Wang J, Gaskov AM. Reduced graphene oxide (rGO) encapsulated Co_3O_4 composite nanofibers for highly selective ammonia sensors. Sensors and Actuators B: Chemical. 2016;**222**:864-870. DOI: 10.1016/j.snb.2015.09.021

[135] Rajaji U, Murugan K, Chen SM, Govindasamy M, Chen TW, Lin PH, et al. Graphene oxide encapsulated 3D porous chalcopyrite ($CuFeS_2$) nanocomposite as an emerging electrocatalyst for agro-hazardous (methyl paraoxon) detection in vegetables. Composites. Part B, Engineering. 2019;**160**:268-276. DOI: 10.1016/j.compositesb.2018.10.042

[136] Myung S, Yin PT, Kim C, Park J, Solanki A, Reyes PI, et al. Label-free polypeptide-based enzyme detection using a graphene-nanoparticle hybrid sensor. Advanced Materials. 2012;**24**:6081-6087. DOI: 10.1002/adma.201202961

[137] Zhou M, Cai T, Pu F, Chen H, Wang Z, Zhang H, et al. Graphene/carbon-coated si nanoparticle hybrids as high-performance anode materials for li-ion batteries. ACS Applied Materials & Interfaces. 2013;**5**:3449-3455. DOI: 10.1021/am400521n

[138] Tanaka Y, Shimizu K, Akashi K, Onda K, Uesugi Y, Ishijima T, et al. High rate synthesis of graphene-encapsulated silicon nanoparticles using

pulse-modulated induction thermal plasmas with intermittent feedstock feeding. Japanese Journal of Applied Physics. 2020;**59**:SHHE07. DOI: 10.35848/1347-4065/ab71db

[139] Ronkainen NJ, Halsall HB, Heineman WR. Electrochemical biosensors. Chemical Society Reviews. 2010;**39**:1747-1763. DOI: 10.1039/b714449k

[140] Pilas J, Selmer T, Keusgen M, Schöning MJ. Screen-printed carbon electrodes modified with graphene oxide for the design of a reagent-free NAD+-dependent biosensor array. Analytical Chemistry. 2019;**91**:15293-15299. DOI: 10.1021/acs.analchem.9b04481

[141] Xu M, Shen Y, Wang L, Gong C, Chen S. A novel H_2O_2 biosensor based on the composite of MP-11 encapasulated in PCN-333 (Al)-graphene oxide. International Journal of Electrochemical Science. 2017;**12**:10390-10401. DOI: 10.20964/2017.11.45

[142] López Marzo AM, Mayorga-Martinez CC, Pumera M. 3D-printed graphene direct electron transfer enzyme biosensors. Biosensors & Bioelectronics. 2020;**151**:111980. DOI: 10.1016/j.bios.2019.111980

[143] Hashim HS, Fen YW, Omar NAS, Daniyal WMEMM, Saleviter S, Abdullah J. Structural, optical and potential sensing properties of tyrosinase immobilized graphene oxide thin film on gold surface. Optik (Stuttg). 2020;**212**:164786. DOI: 10.1016/j.ijleo.2020.164786

[144] Chou JC, Wu CY, Liao YH, Lai CH, Yan SJ, Wu YX, et al. Enzymatic urea sensor based on graphene oxide/titanium dioxide films modified by urease-magnetic beads. IEEE Transactions on Nanotechnology. 2019;**18**:336-344. DOI: 10.1109/TNANO.2019.2907225

[145] Fenoy GE, Marmisollé WA, Azzaroni O, Knoll W. Acetylcholine biosensor based on the electrochemical functionalization of graphene field-effect transistors. Biosensors & Bioelectronics. 2020;**148**:111796. DOI: 10.1016/j.bios.2019.111796

[146] Krishnan SK, Singh E, Singh P, Meyyappan M, Nalwa HS. A review on graphene-based nanocomposites for electrochemical and fluorescent biosensors. RSC Advances. 2019;**9**: 8778-8781. DOI: 10.1039/c8ra09577a

[147] Yao Y, Wang G, Chu G, An X, Guo Y, Sun X. The development of a novel biosensor based on gold nanocages/graphene oxide-chitosan modified acetylcholinesterase for organophosphorus pesticide detection. New Journal of Chemistry. 2019;**43**:13816-13826. DOI: 10.1039/c9nj02556a

[148] Bao J, Huang T, Wang Z, Yang H, Geng X, Xu G, et al. 3D graphene/copper oxide nano-flowers based acetylcholinesterase biosensor for sensitive detection of organophosphate pesticides. Sensors and Actuators B: Chemical. 2019;**279**:95-101. DOI: 10.1016/j.snb.2018.09.118

[149] Zhang J, Wang B, Li Y, Shu W, Hu H, Yang L. An acetylcholinesterase biosensor with high stability and sensitivity based on silver nanowire-graphene-TiO_2 for the detection of organophosphate pesticides. RSC Advances. 2019;**9**:25248-25256. DOI: 10.1039/c9ra02140j

[150] Zhang P, Sun T, Rong S, Zeng D, Yu H, Zhang Z, et al. A sensitive amperometric AChE-biosensor for organophosphate pesticides detection based on conjugated polymer and Ag-rGO-NH_2 nanocomposite. Bioelectrochemistry. 2019;**127**:163-170. DOI: 10.1016/j.bioelechem.2019.02.003

[151] Mogha NK, Sahu V, Sharma M, Sharma RK, Masram DT. Biocompatible ZrO_2- reduced graphene oxide immobilized AChE biosensor for chlorpyrifos detection. Materials and Design. 2016;**111**:312-320. DOI: 10.1016/j.matdes.2016.09.019

[152] Aghaie A, Khanmohammadi A, Hajian A, Schmid U, Bagheri H. Nonenzymatic electrochemical determination of paraoxon ethyl in water and fruits by graphene-based NiFe bimetallic phosphosulfide nanocomposite as a superior sensing layer. Food Analytical Methods. 2019;**12**:1545-1555. DOI: 10.1007/s12161-019-01486-8

[153] Hondred JA, Medintz IL, Claussen JC. Enhanced electrochemical biosensor and supercapacitor with 3D porous architectured graphene via salt impregnated inkjet maskless lithography. Nanoscale Horizons. 2019;**4**:735-746. DOI: 10.1039/c8nh00377g

[154] Dong P, Jiang B, Zheng J. A novel acetylcholinesterase biosensor based on gold nanoparticles obtained by electroless plating on three-dimensional graphene for detecting organophosphorus pesticides in water and vegetable samples. Analytical Methods. 2019;**11**:2428-2434. DOI: 10.1039/c9ay00549h

[155] Yang L, Wang G, Liu Y. An acetylcholinesterase biosensor based on platinum nanoparticles- carboxylic graphene-nafion-modified electrode for detection of pesticides. Analytical Biochemistry. 2013;**437**:144-149. DOI: 10.1016/j.ab.2013.03.004

[156] Yang L, Wang G, Liu Y, Wang M. Development of a biosensor based on immobilization of acetylcholinesterase on NiO nanoparticles–carboxylic graphene–nafion modified electrode for detection of pesticides. Talanta. 2013;**113**:135-141. DOI: 10.1016/j.talanta.2013.03.025

[157] Yang L, Zhou Q, Wang G, Yang Y. Acetylcholinesterase biosensor based on SnO_2 nanoparticles-carboxylic graphene-nafion modified electrode for detection of pesticides. Biosensors & Bioelectronics. 2013;**49**:25-31. DOI: 10.1016/j.bios.2013.04.037

[158] Jeyapragasam T, Saraswathi R, Chen SM, Chen TW. Acetylcholinesterase biosensor for the detection of methyl parathion at an electrochemically reduced graphene oxide-nafion modified glassy carbon electrode. International Journal of Electrochemical Science. 2017;**12**: 4768-4781. DOI: 10.20964/2017.06.77

[159] Tan X, Hu Q, Wu J, Li X, Li P, Yu H, et al. Electrochemical sensor based on molecularly imprinted polymer reduced graphene oxide and gold nanoparticles modified electrode for detection of carbofuran. Sensors and Actuators B: Chemical. 2015;**220**:216-221. DOI: 10.1016/j.snb.2015.05.048

[160] Zhang M, Zhao HT, Xie TJ, Yang X, Dong AJ, Zhang H, et al. Molecularly imprinted polymer on graphene surface for selective and sensitive electrochemical sensing imidacloprid. Sensors and Actuators B: Chemical. 2017;**252**:991-1002. DOI: 10.1016/j.snb.2017.04.159

[161] Chao M, Chen M. Electrochemical determination of Phoxim in food samples employing a Graphene-modified glassy carbon electrode. Food Analytical Methods. 2014;**7**:1729-1736. DOI: 10.1007/s12161-014-9813-y

[162] Wu L, Lei W, Han Z, Zhang Y, Xia M, Hao Q. A novel non-enzyme amperometric platform based on poly(3-methylthiophene)/nitrogen

doped graphene modified electrode for determination of trace amounts of pesticide phoxim. Sensors and Actuators B: Chemical. 2015;**206**:495-501. DOI: 10.1016/j.snb.2014.09.098

[163] Zheng Y, Wang A, Lin H, Fu L, Cai W. A sensitive electrochemical sensor for direct phoxim detection based on an electrodeposited reduced graphene oxide-gold nanocomposite. RSC Advances. 2015;**5**:15425-15430. DOI: 10.1039/c4ra15872e

[164] Karimian N, Fakhri H, Amidi S, Hajian A, Arduini F, Bagheri H. A novel sensing layer based on metal-organic framework UiO-66 modified with TiO_2-graphene oxide: Application to rapid, sensitive and simultaneous determination of paraoxon and chlorpyrifos. New Journal of Chemistry. 2019;**43**:2600-2609. DOI: 10.1039/c8nj06208k

[165] Liu Q, Fei A, Huan J, Mao H, Wang K. Effective amperometric biosensor for carbaryl detection based on covalent immobilization acetylcholinesterase on multiwall carbon nanotubes/graphene oxide nanoribbons nanostructure. Journal of Electroanalytical Chemistry. 2015;**740**:8-13. DOI: 10.1016/j.jelechem.2014.12.037

[166] Liu Y, Wang G, Li C, Zhou Q, Wang M, Yang L. A novel acetylcholinesterase biosensor based on carboxylic graphene coated with silver nanoparticles for pesticide detection. Materials Science and Engineering: C. 2014;**35**:253-258. DOI: 10.1016/j.msec.2013.10.036

[167] Wang G, Tan X, Zhou Q, Liu Y, Wang M, Yang L. Synthesis of highly dispersed zinc oxide nanoparticles on carboxylic graphene for development a sensitive acetylcholinesterase biosensor. Sensors and Actuators B: Chemical. 2014;**190**:730-736. DOI: 10.1016/j.snb.2013.09.042

[168] Zheng Y, Liu Z, Jing Y, Li J, Zhan H. An acetylcholinesterase biosensor based on ionic liquid functionalized graphene-gelatin-modified electrode for sensitive detection of pesticides. Sensors and Actuators B: Chemical. 2015;**210**:389-397. DOI: 10.1016/j.snb.2015.01.003

[169] Zhang L, Zhang A, Du D, Lin Y. Biosensor based on Prussian blue nanocubes/reduced graphene oxide nanocomposite for detection of organophosphorus pesticides. Nanoscale. 2012;**4**:4674-4679. DOI: 10.1039/c2nr30976a

[170] Zhao H, Ji X, Wang B, Wang N, Li X, Ni R, et al. An ultra-sensitive acetylcholinesterase biosensor based on reduced graphene oxide-Au nanoparticles-β-cyclodextrin/Prussian blue-chitosan nanocomposites for organophosphorus pesticides detection. Biosensors & Bioelectronics. 2015;**65**:23-30. DOI: 10.1016/j.bios.2014.10.007

[171] Ensafi AA, Noroozi R, Zandi-Atashbar N, Rezaei B. Cerium (IV) oxide decorated on reduced graphene oxide, a selective and sensitive electrochemical sensor for fenitrothion determination. Sensors actuators B Chem. 2017;**245**:980-987. DOI: 10.1016/j.snb.2017.01.186

[172] Shams N, Lim HN, Hajian R, Yusof NA, Abdullah J, Sulaiman Y, et al. A promising electrochemical sensor based on Au nanoparticles decorated reduced graphene oxide for selective detection of herbicide diuron in natural waters. Journal of Applied Electrochemistry. 2016;**46**:655-666. DOI: 10.1007/s10800-016-0950-4

[173] Yang Y, Asiri AM, Du D, Lin Y. Acetylcholinesterase biosensor based on a gold nanoparticle-polypyrrole-reduced graphene oxide nanocomposite modified electrode for the amperometric detection of organophosphorus pesticides. Analyst.

2014;**139**:3055-3060. DOI: 10.1039/c4an00068d

[174] Li Y, Zhang Y, Han G, Xiao Y, Li M, Zhou W. An acetylcholinesterase biosensor based on graphene/polyaniline composite film for detection of pesticides. Chinese Journal of Chemistry. 2016;**34**:82-88. DOI: 10.1002/cjoc.201500747

[175] Li Y, Shi L, Han G, Xiao Y, Zhou W. Electrochemical biosensing of carbaryl based on acetylcholinesterase immobilized onto electrochemically inducing porous graphene oxide network. Sensors and Actuators B: Chemical. 2017;**238**:945-953. DOI: 10.1016/j.snb.2016.07.152

[176] Bao J, Hou C, Chen M, Li J, Huo D, Yang M, et al. Plant esterase-chitosan/gold nanoparticles-graphene nanosheet composite-based biosensor for the ultrasensitive detection of organophosphate pesticides. Journal of Agricultural and Food Chemistry. 2015;**63**:10319-10326. DOI: 10.1021/acs.jafc.5b03971